MW01670530

# Brooks/Cole

# One-Unit Series

# in Precalculus Mathematics

# GEOMETRY

Brooks/Cole

One-Unit Series

in Precalculus Mathematics

# GEOMETRY

Karl J. Smith

BROOKS/COLE PUBLISHING COMPANY

Pacific Grove, California

Brooks/Cole Publishing Company
A Division of Wadsworth, Inc.



Printed in the United States of America

10 9 8 7 6 5 4 3 2

Library of Congress Cataloging-in-Publication Data

Smith, Karl J.
Geometry/ Karl J. Smith
p. cm. -- (Brooks/Cole one-unit series in precalculus mathematics)
Includes index.
ISBN 0-534-14929-4 (pbk.)
1. Geometry. I. Title II. Series
QA445.S58 1991
516--dc20

90-24931
CIP

Sponsoring Editor: Paula-Christy Heighton
Editorial Assistant: Elaine Giuliano
Marketing Representative: Deborah Mobley
Production Editor: Joan Marsh
Production Service: Bookman Productions
Interior Design: Karl J. Smith
Cover Design: Vernon T. Boes
Typesetting: Karlin Enterprises
Cover Printing: Malloy Lithographing, Inc.
Printing and Binding: Malloy Lithographing, Inc.

# PREFACE

Although many courses in the college curriculum do not require a complete course in geometry, they do require a knowledge of several geometrical concepts and of the associated terminology. In particular, students preparing for the TASP test in Texas and the ELM test in California often need supplementary work on the geometric concepts included in this supplement.

This material can be used as a supplement for an existing course, as a short (one-unit) course, or as a basis for independent study, which is why the answers to all the problems are provided in an appendix. You will find the problems in this book divided into A and B parts. The A problems follow the examples and text exposition rather closely, while the B problems require that the student *understand and use* the ideas of the section.

I would like to extend my thanks to the following reviewers for their comments and suggestions: Kent Bentley, California State University, Fullerton; Karen Emerson, St. Petersburg Junior College; Jane Edger, Brevard Community College; Carroll Wells, Western Kentucky University; Beth Hooper, Golden West College; and Dawson Carr, Sandhills Community College.

In particular, my appreciation to Carol Kublin for checking all of the examples and answers for accuracy.

Karl J. Smith
Sebastopol, CA

# CONTENTS

# CHAPTER 1 ANGLES AND POLYGONS

### 1.1 Geometry

Geometry, or "earth measure," was one of the first branches of mathematics. Both the Egyptians and the Babylonians needed geometry for construction, land measurement, and commerce.

Geometry involves **points** as well as sets of points called **lines, planes,** and **surfaces.** Certain concepts in geometry are called **undefined terms.** For example, what is a line? You might say, "I know what a line is!" But try to define a line. Is it a set of points? Any set of points? What is a point? You might say, "A point is something that has no length, width, or thickness." Or, "A point is a location in space." Certainly these are not satisfactory definitions, because they involve other terms that are not defined. We will therefore take the terms *point, line,* and *plane* as undefined terms.

### 1.2 Polygons

A polygon is a geometric figure in a plane that consists of three or more straight-line segments that enclose a region of the plane. Polygons are classified according to their number of sides, as shown in Figure 1.

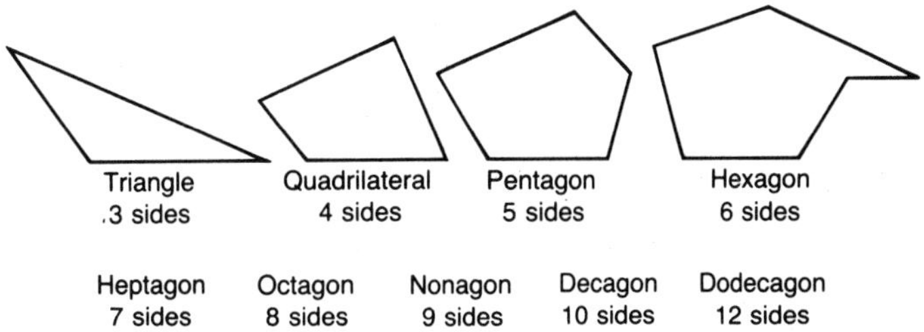


Figure 1

### 1.3 Angles

A connecting point of two sides of a polygon is called a **vertex** (plural **vertices**) and is usually designated by a capital letter. The **angles** between the sides are sometimes also denoted by a capital letter, but other ways of denoting angles are shown in Figure 2.

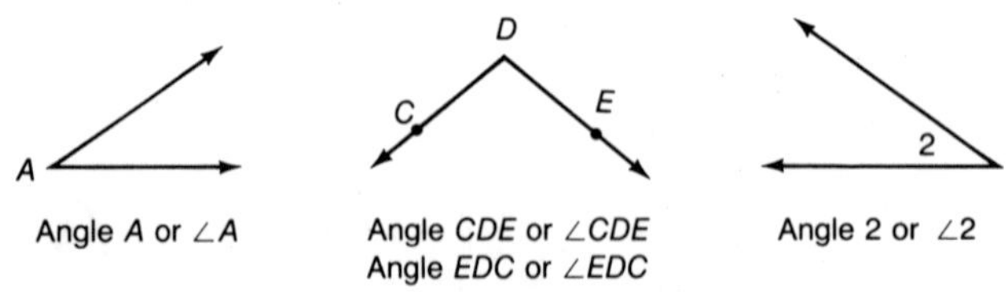


Figure 2

Two angles are said to be **congruent** if they have the same measure. If we write $m$ in front of an angle symbol, we mean the measure of the angle rather than the angle itself. Angles are usually measured using a unit called a **degree,** which is defined to be $\frac{1}{360}$ of a full revolution. The symbol ° is used to designate degrees. To measure an angle, you can use a **protractor,** but in this book the angles whose measures we need will be labeled as in Figure 3.

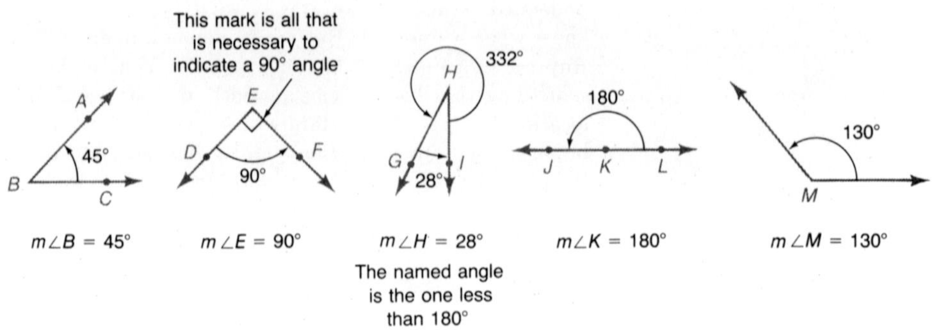


Figure 3

Angles with measures greater than zero are sometimes classified according to their measures, as shown in Table 1.1.

**Table 1.1**
Classification of Angles

| Type | Classification |
|---|---|
| Angles with measure less than 90° | Acute |
| Angles with measure equal to 90° | Right |
| Angles with measure between 90° and 180° | Obtuse |
| Angles with measure equal to 180° | Straight |
| **Pairs of angles** | |
| Sum of measures is 90° | Complementary |
| Sum of measures is 180° | Supplementary |

**EXAMPLE 1** Label the angles in Figure 1.1 by classification.

*Solution* $\angle B$ is acute; $\angle E$ is right; $\angle H$ is acute; $\angle K$ is straight; $\angle M$ is obtuse. □

### 1.4 Angles with Parallel Lines

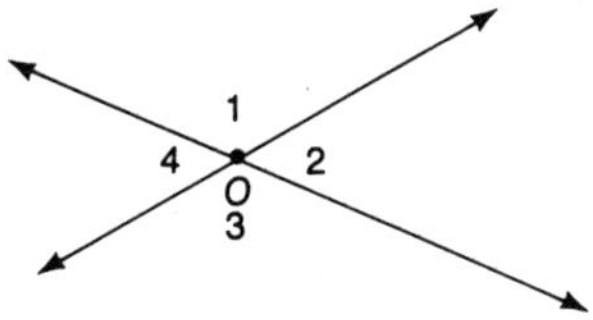


Figure 4

Consider any two distinct (different) intersecting lines in a plane, and let $O$ be the point of intersection, as shown in Figure 4. These lines must form four angles. Angles that have a common vertex and lie on opposite sides of a common side are called **adjacent angles**; nonadjacent angles are called **vertical angles.**

**EXAMPLE 2**

**a.** Name the adjacent angles in Figure 4.

**b.** Name the vertical angles in Figure 4.

*Solution*

The pairs are $\angle 1$ and $\angle 2$; $\angle 2$ and $\angle 3$; $\angle 3$ and $\angle 4$; $\angle 4$ and $\angle 1$.

The pairs are $\angle 1$ and $\angle 3$; $\angle 2$ and $\angle 4$. □

Now consider any three distinct lines such that exactly two of them — say $\ell_1$ and $\ell_2$ — are parallel (lie in the same plane and never intersect), and also that the third line $\ell_3$ intersects the parallel lines at points $P$ and $Q$, as shown in Figure 5.

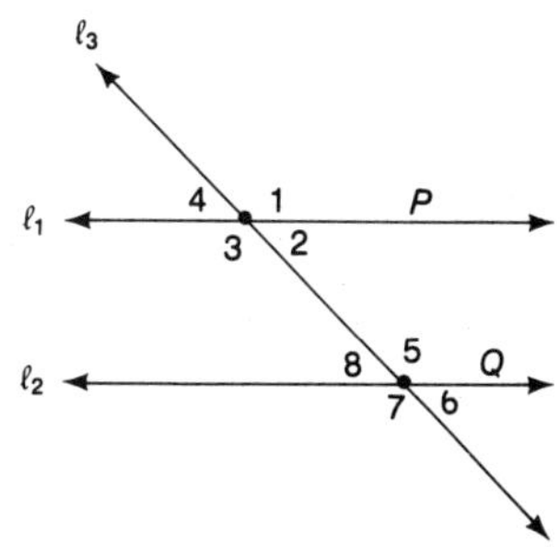


Figure 5

The line $\ell_3$ is called a **transversal.** Notice that several angles are formed:

**Vertical angles** are congruent.

**Alternate interior angles** are nonadjacent pairs of angles whose interiors lie between the parallel lines, but on opposite sides of the transversal. Alternate interior angles are congruent.

**Alternate exterior angles** are nonadjacent pairs of angles that lie outside the parallel lines, but on opposite sides of the transversal. Alternate exterior angles are congruent.

**Corresponding angles** are two nonadjacent angles whose interiors lie on the same side of the transversal such that one angle lies between the parallel lines and the other side lies on the outside. Corresponding angles are congruent.

**EXAMPLE 3**

**a.** Name the vertical angles in Figure 5.

**b.** Name the alternate interior angles in Figure 5.

c. Name the alternate exterior angles in Figure 5.

d. Name the corresponding angles in Figure 5.

*Solution*

a. $\angle 1$ and $\angle 3$; $\angle 2$ and $\angle 4$; $\angle 5$ and $\angle 7$; and $\angle 6$ and $\angle 8$.

b. $\angle 2$ and $\angle 8$; $\angle 3$ and $\angle 5$.

c. $\angle 1$ and $\angle 7$; $\angle 4$ and $\angle 6$.

d. $\angle 1$ and $\angle 5$; $\angle 2$ and $\angle 6$; $\angle 3$ and $\angle 7$; $\angle 4$ and $\angle 8$. □

Summarizing the results from Example 3, we can say that the following angles are congruent (written $\simeq$):

$$\angle 1 \simeq \angle 3 \simeq \angle 5 \simeq \angle 7 \quad \text{and also} \quad \angle 2 \simeq \angle 4 \simeq \angle 6 \simeq \angle 8$$

Moreover, the pairs of adjacent angles are all supplementary (their measures add up to 180°).

**EXAMPLE 4** Find the measures of all the angles in Figure 6, where $\ell_1$ and $\ell_2$ are parallel.

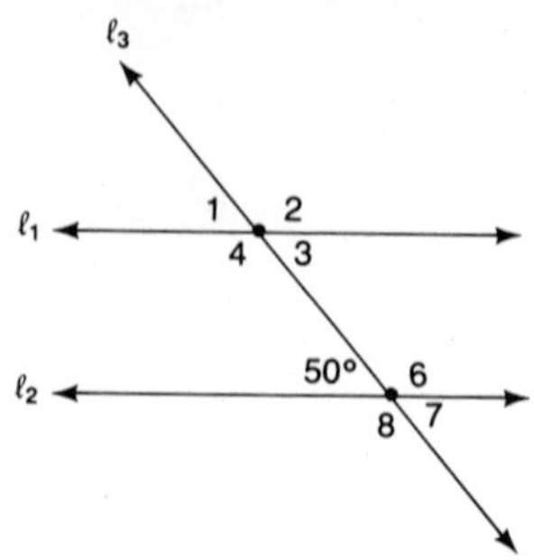


Figure 6

*Solution*

| | |
|---|---|
| $m\angle 1 = 50°$ | *Corresponding angles* |
| $m\angle 2 = 180° - 50° = 130°$ | *$\angle 1$ and $\angle 2$ are supplementary* |
| $m\angle 3 = 50°$ | *$\angle 3$ and $\angle 1$ are vertical* |
| $m\angle 4 = 130°$ | *Vertical angles* |
| $m\angle 6 = 130°$ | *Supplementary angles* |
| $m\angle 7 = 50°$ | *Vertical angles* |
| $m\angle 8 = 130°$ | *Supplementary angles* |

□

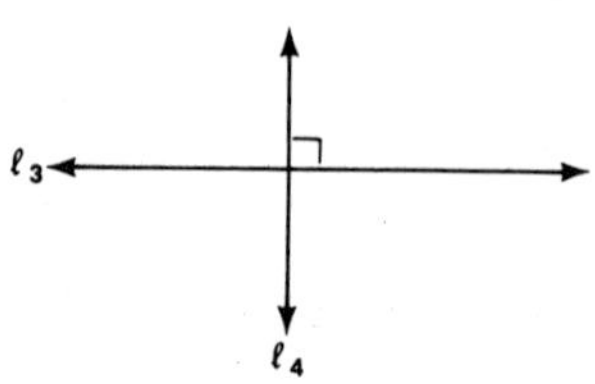


Figure 7

### 1.5 Perpendicular Lines

If two lines intersect so that the adjacent angles are congruent, then the lines are **perpendicular**. Simply, lines that intersect to form angles of 90° (right angles) are called *perpendicular lines*. In Figure 7, lines $\ell_3$ and $\ell_4$ intersect to form a right angle and therefore are perpendicular lines.

In a diagram on a printed page, any line that is parallel to the top and bottom edge of the page is considered a **horizontal line.** Lines that are perpendicular to a horizontal line or to the top and bottom of the page are considered **vertical lines.** In Figure 8, line $\ell_5$ is a horizontal line and line $\ell_6$ is a vertical line.

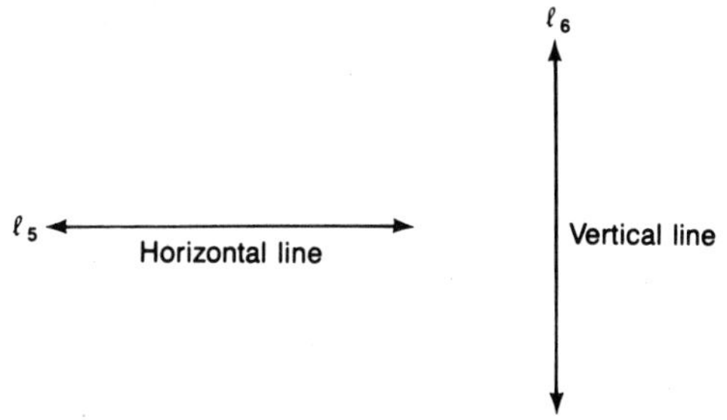


Figure 8

**EXAMPLE 5**

The diagram in Figure 9 shows lines in the same plane. Which of the following statements are true?

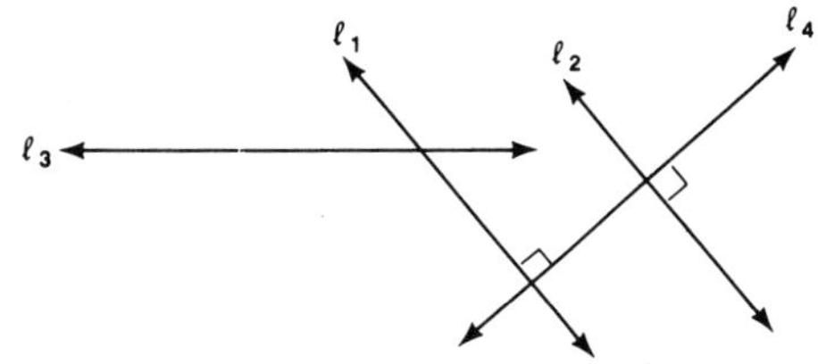


Figure 9

A. Lines $\ell_1$ and $\ell_2$ are parallel and horizontal.
B. Lines $\ell_4$ and $\ell_2$ are intersecting but not perpendicular.
C. Lines $\ell_2$ and $\ell_3$ are intersecting lines.
D. None of these statements are true.

*Solution*

Lines $\ell_1$ and $\ell_2$ are parallel, but they are not horizontal. Therefore, A is false. Lines $\ell_4$ and $\ell_2$ are both intersecting and perpendicular. Therefore, B is false. Lines $\ell_2$ and $\ell_3$ are intersecting since the arrows on $\ell_3$ indicate that it will intersect line $\ell_2$. Therefore, C is true and your choice is C. □

The following list identifies some important statements about parallel and perpendicular lines that are often used for solving geometry problems.

1. If a line is perpendicular to one of two parallel lines in a plane, then it is also perpendicular to the other line.
2. If two lines intersect to form congruent adjacent angles, then the lines are perpendicular.
3. If two parallel lines are intersected by a transversal, then each pair of alternate interior angles is congruent.
4. If a transversal intersects a pair of lines in such a way that the alternate interior angles are congruent, then the two lines intersected by the transversal are parallel.
5. If two pairs of parallel lines are intersected by a transversal, then each pair of alternate exterior angles is congruent.
6. If a transversal intersects a pair of lines in such a way that the alternate exterior angles are congruent, then the two lines

intersected by the transversal are parallel.

7. If two parallel lines are intersected by a transversal, then each pair of corresponding angles is congruent.
8. If a transversal intersects a pair of lines in such a way that the corresponding angles are congruent, then the two lines intersected by the transversal are parallel.
9. If two parallel lines are intersected by a transversal, then the exterior angles on the same side of the transversal are supplementary.
10. If a transversal intersects a pair of lines in such a way that the exterior angles on the same side of a transversal are supplementary, then the two lines intersected by the transversal are parallel.
11. If two parallel lines are intersected by a transversal, then the interior angles on the same side of the transversal are supplementary.
12. If a transversal intersects a pair of lines in such a way that the interior angles on the same side of the transversal are supplementary, then the two lines intersected by the transversal are parallel.

## 1.6 Triangles

### Classification

One of the most frequently encountered polygons is the triangle. Every triangle has six parts: three sides and three angles. We name the sides after the endpoints of the line segments, and we name the angles after the vertices (see Figure 10).

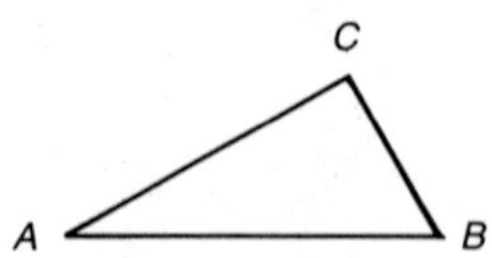


| Sides | Angles |
|---|---|
| $AC$ | $\angle A$ |
| $CB$ | $\angle B$ |
| $AB$ | $\angle C$ |

Figure 10

Triangles are often classified by sides or by angles:

*By sides:*

*Scalene* – no equal sides
*Isosceles* – at least two equal sides
*Equilateral* – three equal sides

*By angles:*

*Acute* – three acute angles
*Right* – one right angle
*Obtuse* – one obtuse angle

### Sum of the measures of angles

One of the most basic properties of triangles involves the sum of the measures of the angles of a triangle. To discover this property for yourself, place a pencil with an eraser along one side of any triangle, as shown in Figure 11a. Now rotate the pencil to correspond to the size of $\angle A$ as shown in Figure 11b. As you can see, your pencil is along side $AC$. Next, rotate the pencil through $\angle C$, as shown in Figure 11c. Finally, rotate the pencil through $\angle B$. Notice that the pencil has been rotated the same amount as the sum of the angles of the triangle. Also notice that the pencil's orientation is exactly reversed from its starting position.

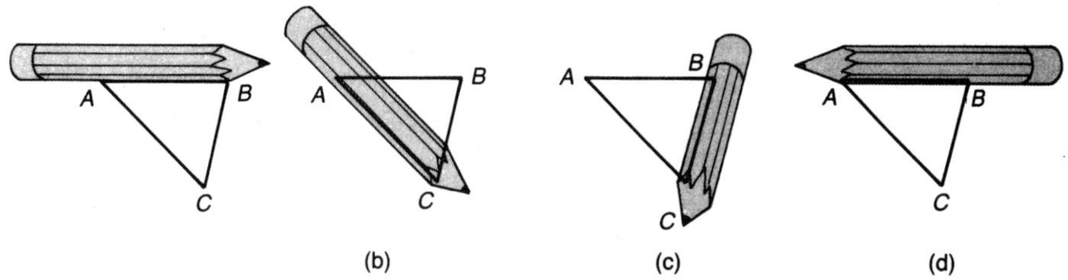


Figure 11 Demonstration that the sum of the measures of angles in a triangle is 180°

This leads us to the following important property.

**SUM OF THE MEASURES OF ANGLES IN A TRIANGLE**

> The sum of the measures of the angles in any triangle is 180°.

**EXAMPLE 6** Find the missing angle measure in the triangle in Figure 12.

*Solution* Let $x$ represent the missing angle measure.

$$\begin{aligned} 65 + 82 + x &= 180 \\ 147 + x &= 180 \\ x &= 33 \end{aligned}$$

The missing angle's measure is 33°.

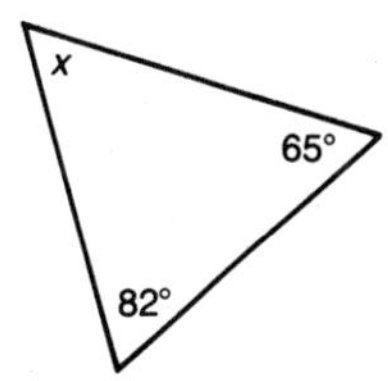


Figure 12 □

**EXAMPLE 7** Find the measures of the angles of a triangle if it is known that the measures are $x$, $2x - 15$, and $3(x + 17)$ degrees.

*Solution* Using the property that the sum of the measures of angles in a triangle is 180°, we have

$$\begin{aligned} x + (2x - 15) + 3(x + 17) &= 180 \\ x + 2x - 15 + 3x + 51 &= 180 && \textit{Eliminate parentheses} \\ 6x + 36 &= 180 && \textit{Combine similar terms} \\ 6x &= 144 && \textit{Subtract 36 from both sides} \\ x &= 24 && \textit{Divide both sides by 6} \end{aligned}$$

Now, we find the angle measures:

$$\begin{aligned} x &= 24 \\ 2x - 15 &= 2(24) - 15 = 33 \\ 3(x + 17) &= 3(24 + 17) = 123 \end{aligned}$$

□

**EXAMPLE 8** In $\triangle ABC$ shown at right, if $m\angle BAC = 120°$, what is $x$?

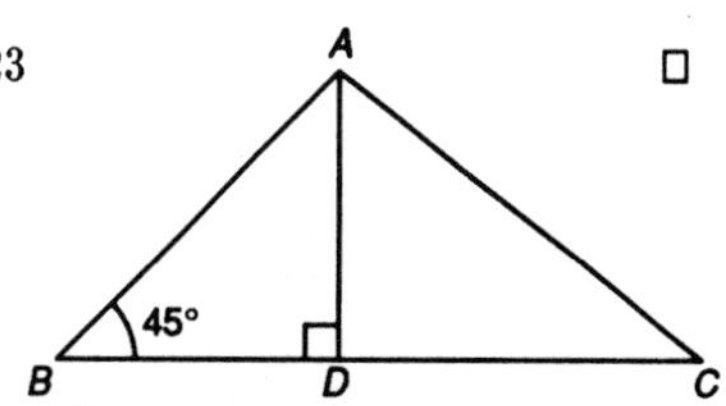


*Solution* Since the sum of the angles of a triangle is 180°, we see that

$$m\angle A + m\angle B + m\angle C = 180^\circ$$
$$120^\circ + 45^\circ + m\angle C = 180^\circ$$
$$m\angle C = 15^\circ$$

Now $\angle ADC = 90^\circ$, so again using the same property on $\triangle ADC$,

$$m\angle ADC + m\angle C + x = 180^\circ$$
$$90^\circ + 15^\circ + x = 180^\circ$$
$$x = 75^\circ$$

□

### Exterior angles

An **exterior angle** of a triangle is the angle on the other side of an extension of one side of the triangle. An example is the angle whose measure is marked as $x$ in Figure 13.

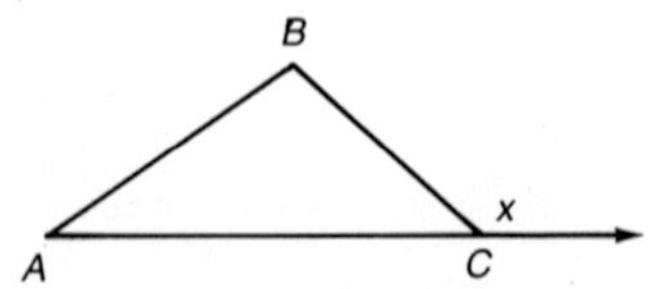


Figure 13

Notice that the following relationships are true for any $\triangle ABC$ with exterior angle $x$:

$$m\angle A + m\angle B + m\angle C = 180^\circ \quad \text{and} \quad m\angle C + x = 180^\circ$$

Thus,

$$m\angle A + m\angle B + m\angle C = m\angle C + x = 180^\circ$$
$$m\angle A + m\angle B = x \qquad \textit{Subtract } m\angle C \textit{ from both sides}$$

**EXTERIOR ANGLES OF A TRIANGLE**

> The measure of the exterior angle of a triangle equals the sum of the measures of the two opposite interior angles.

**EXAMPLE 9** Find the value of $x$ in Figure 14.

*Solution*

SUM OF INTERIOR ANGLES = EXTERIOR ANGLES

$$63^\circ + 42^\circ = x$$
$$105^\circ = x$$

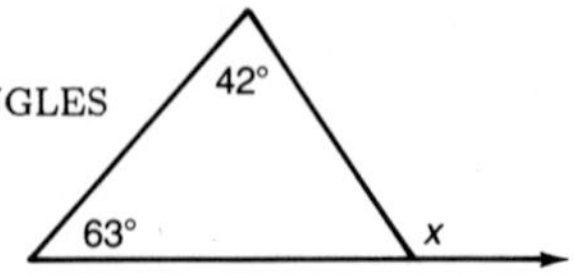


Figure 14

□

### 1.7 Quadrilaterals

Like the triangle, the quadrilateral is a commonly encountered polygon. The first of its properties we will consider involves the sum of the measures of the angles in any quadrilateral.

**ANGLES OF A QUADRILATERAL**

> The sum of the measures of the angles of any quadrilateral is 360°.

This property is easy to show by drawing a diagonal of the quadrilateral, as shown in Figure 15. The result of drawing a diagonal of a quadrilateral is to form two triangles. Since the sum of the measures of angles of a triangle is 180°, we see that the sum of the six angles formed in Figure 15 is

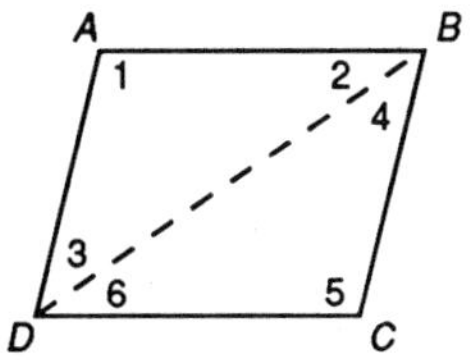

Figure 15

$$m\angle 1 + m\angle 2 + m\angle 3 + m\angle 4 + m\angle 5 + m\angle 6 = 180^\circ + 180^\circ = 360^\circ$$

**Classification**

There are three special quadrilaterals with which you should be familiar. The first is a **parallelogram** — a quadrilateral in which opposite sides are parallel, as shown in Figure 16.

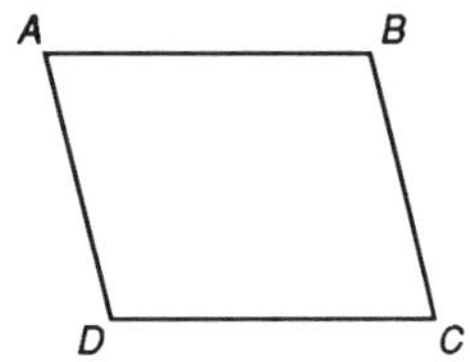

Figure 16 Parallelogram

The words *adjacent* and *opposite* are used to describe both the sides and the angles of a parallelogram. Sides are adjacent if they share a common endpoint, and angles are adjacent if they share a common side. Sides of a parallelogram that are not adjacent are called opposite sides and angles that are not adjacent are called opposite angles.

**ADJACENT ANGLES IN A PARALLELOGRAM**

> In a parallelogram, adjacent angles are supplementary, and opposite angles have equal measures.

Since the opposite sides of a parallelogram are parallel (see Figure 16), $\angle B$ and $\angle 1$ are corresponding angles, so they have the same measure. Since $\angle 1$ and $\angle A$ are obviously supplementary (they form a straight line), it follows that $\angle A$ and $\angle B$ are supplementary. Showing that the opposite angles have equal measures is left as an exercise.

A second type of parallelogram is one in which each angle is a right angle. Such a parallelogram is called a **rectangle.** If the rectangle has sides of the same length, it is called a **square.** Since squares are rectangles, and rectangles are parallelograms, it follows that both squares and rectangles have opposite sides that are parallel.

## 1.8 PROBLEM SET 1

**A** *Name the polygons in Problems 1–12 according to the classifications shown in Figure 1.*

1. 

2. 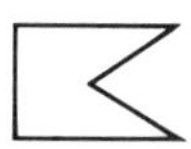

3. 

**4.** 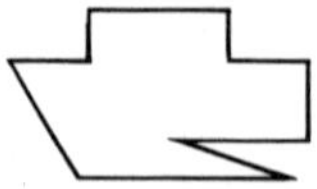

**5.** 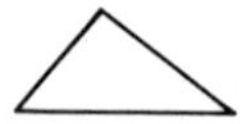

**6.** 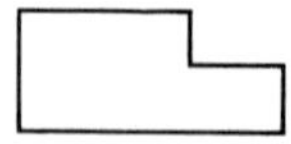

**7.** 

**8.** 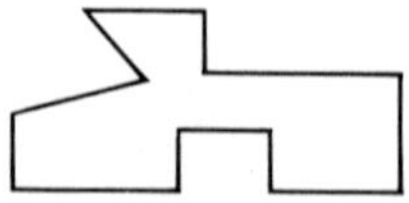

**9.** 

**10.** 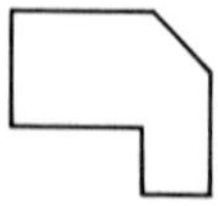

**11.** 

**12.** 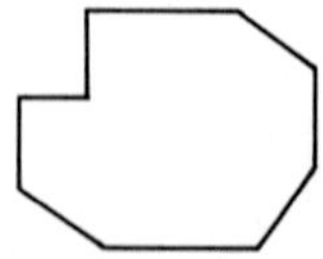

*Consider the following figure for Problems 13–16.*

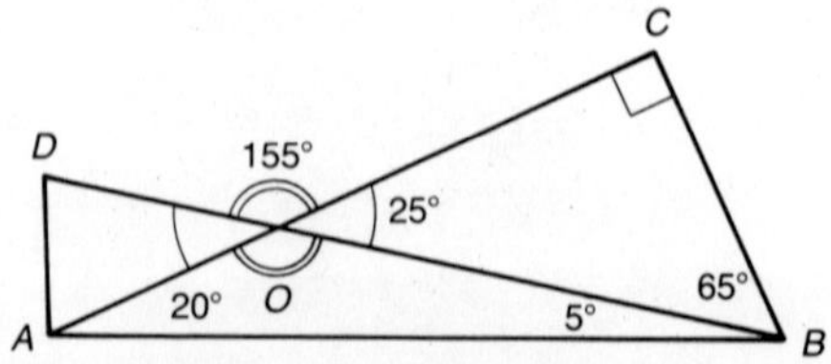


**13.** Classify the angles as acute, right, straight, or obtuse.
**a.** $\angle DOC$ **b.** $\angle AOB$ **c.** $\angle DBC$ **d.** $\angle CAB$ **e.** $\angle DOB$

**14.** Classify the angles as acute, right, straight, or obtuse.
**a.** $\angle C$ **b.** $\angle COB$ **c.** $\angle AOC$ **d.** $\angle DOA$ **e.** $\angle OBA$

**15.** Name a pair of congruent angles.

**16.** Name an angle congruent to $\angle DOB$.

*Consider the following figure for Problems 17–20.*

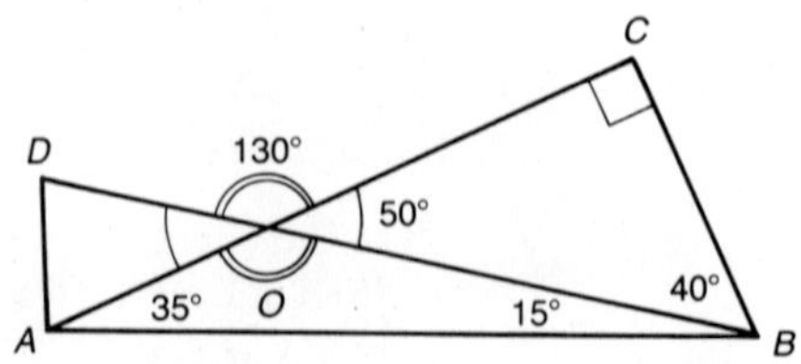


**17.** Classify the angles as acute, right, straight, or obtuse.
**a.** $\angle DOC$ **b.** $\angle AOB$ **c.** $\angle DBC$ **d.** $\angle CAB$ **e.** $\angle DOB$

**18.** Classify the angles as acute, right, straight, or obtuse.
**a.** $\angle C$ **b.** $\angle COB$ **c.** $\angle AOC$ **d.** $\angle DOA$ **e.** $\angle OBA$

**19.** Name a pair of congruent angles.

**20.** Name an angle congruent to $\angle DOB$.

*In Problems 21–26, classify the requested angles shown in the following figure, given that $\ell_1$ and $\ell_2$ are parallel.*

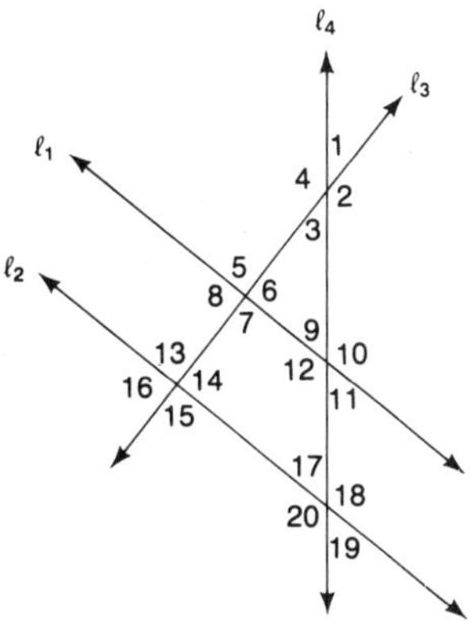


**21.** $m\angle 3$ if $m\angle 1 = 30°$ **22.** $m\angle 6$ if $m\angle 5 = 90°$
**23.** $m\ \angle 17$ if $m\angle 18 = 105°$ **24.** $m\angle 10$ if $m\angle 11 = 90°$
**25.** $m\angle 17$ if $m\angle 11 = 40°$ **26.** $m\angle 7$ if $m\angle 15 = 80°$

*In Problems 27–32, classify the pairs of angles, as shown in the figure for Problems 21–26.*

**27.** $\angle 2$ and $\angle 4$ **28.** $\angle 13$ and $\angle 14$ **29.** $\angle 8$ and $\angle 14$
**30.** $\angle 9$ and $\angle 17$ **31.** $\angle 9$ and $\angle 11$ **32.** $\angle 12$ and $\angle 18$

*Each of the four diagrams below shows lines in the same plane. State whether each of the statements in Problems 33–38 is true or false.*

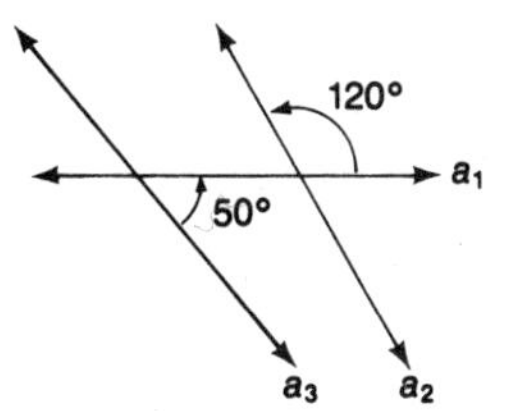


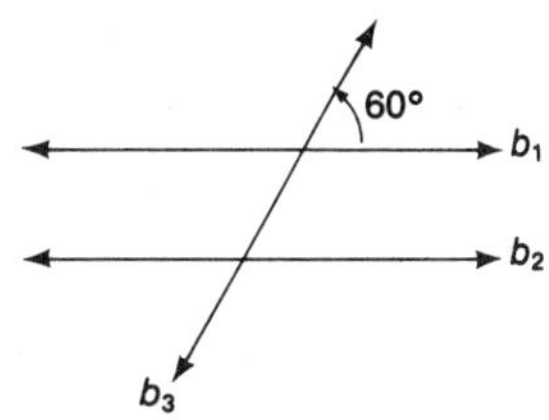


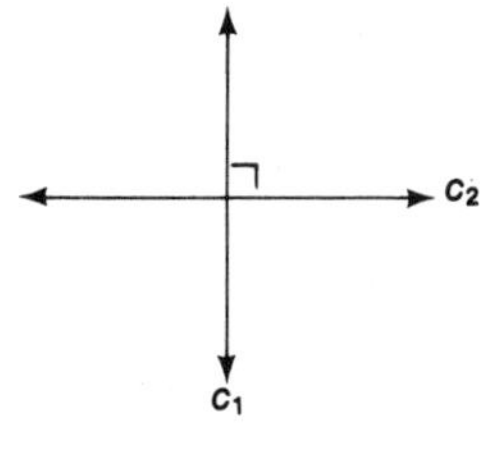


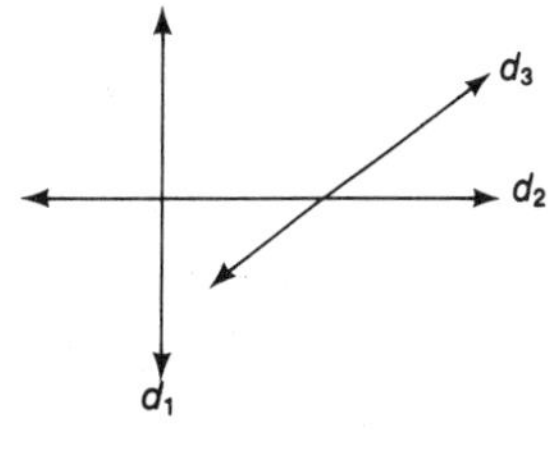


**33.** The lines $a_1$ and $a_3$ are both intersecting and parallel.
**34.** The lines $b_1$ and $b_2$ are vertical.
**35.** The lines $d_1$ and $d_3$ are intersecting.
**36.** The lines $c_1$ and $c_2$ are not intersecting.
**37.** The lines $c_2$ and $d_2$ are horizontal.
**38.** The lines $a_2$ and $a_3$ are both vertical and parallel.
**39.** Find the measures of all the angles in Figure 5 if $m\angle 7 = 110°$.
**40.** Find the measures of all the angles in Figure 5 if $m\angle 2 = 65°$.
**41.** Find the measures of all the angles in Figure 5 if $m\angle 6 = 19°$.

42. Find the measures of all the angles in Figure 5 if $m\angle 1 = 153°$.

**B**

43. Use the figure at right to find $m\angle 1$ if $\ell_1$ is parallel to $\ell_2$ and $m\angle 7 = 68°$.

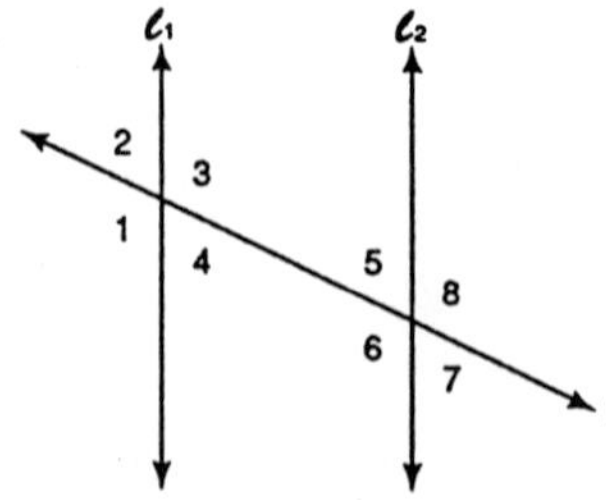


44. In the figure shown below, what is $x$ if $\ell_1$ and $\ell_2$ are parallel and $\ell_3$ is perpendicular to $\ell_1$?

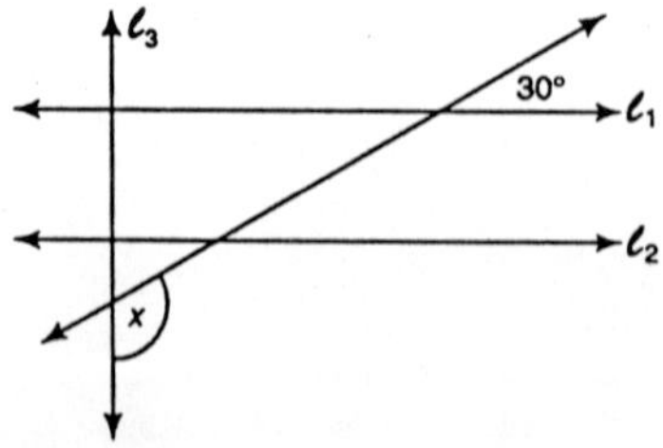


45. Use the figure at right to find $m\angle BCD$ if $AB$ is parallel to $CD$.

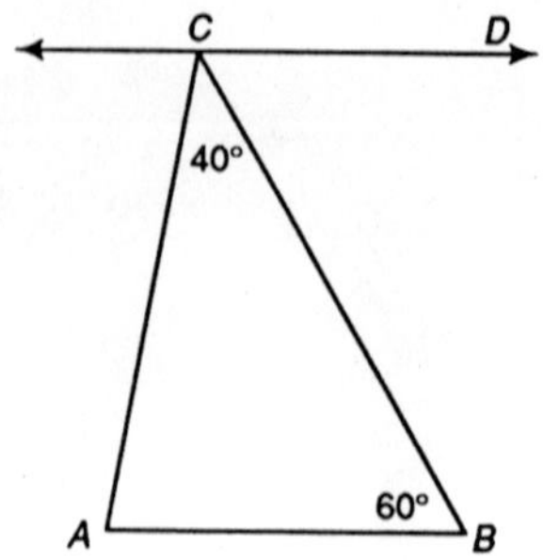


46. In square $ABCD$ shown below, find $x$ if $m\angle AEF = 140°$.

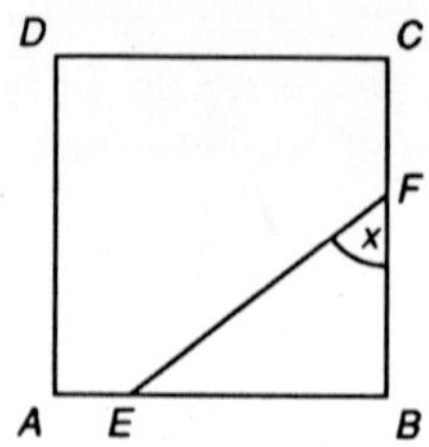


*In Problems 47–52, find the third angle in each triangle.*

47.

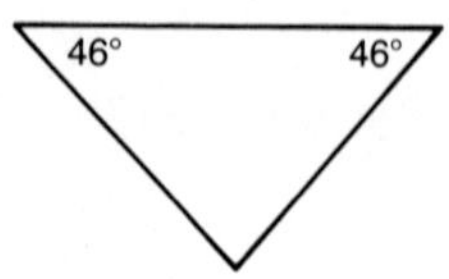


48.

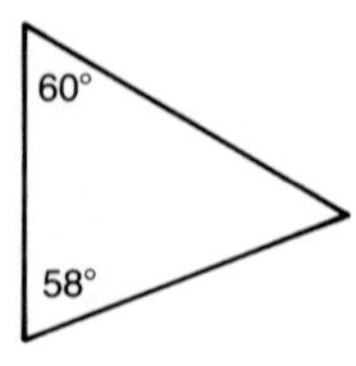


49.

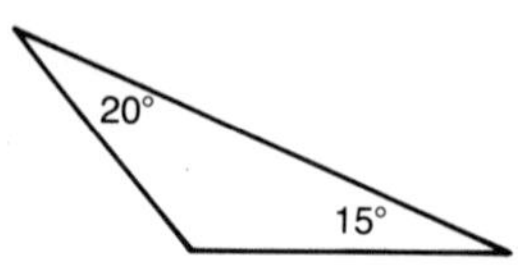


50.

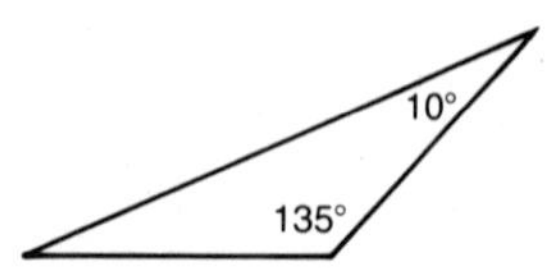


51.

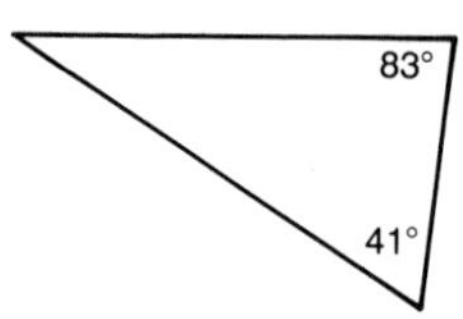


52.

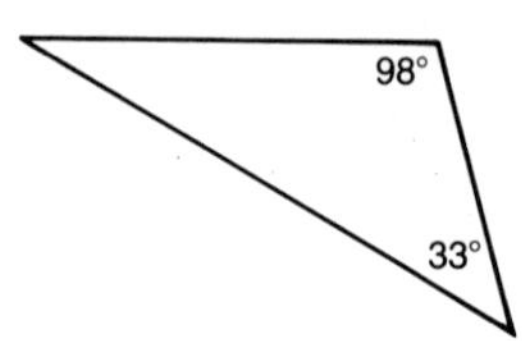

*Find the value of $x$ in each of the triangles in Problems 53–64.*

**53.**

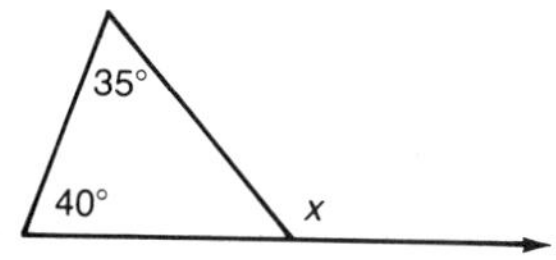


**54.**

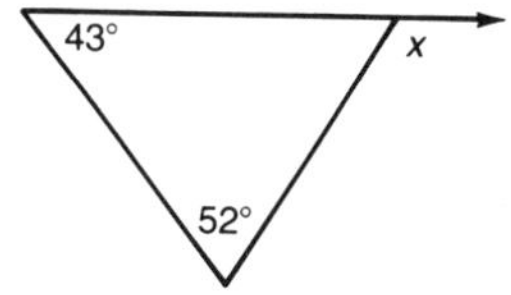


**55.**

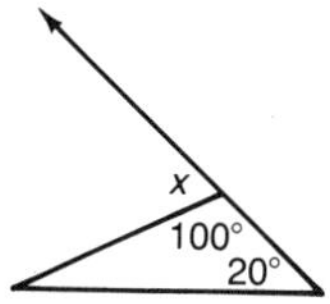


**56.**

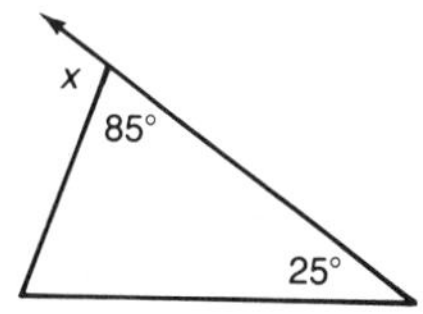


**57.**

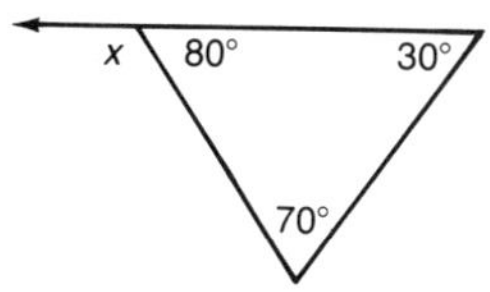


**58.**

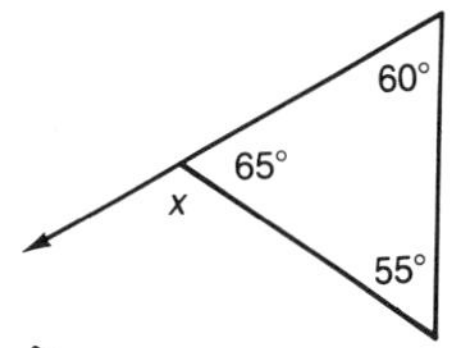


**59.**

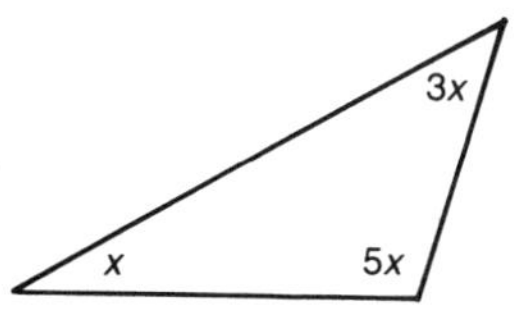


**60.**

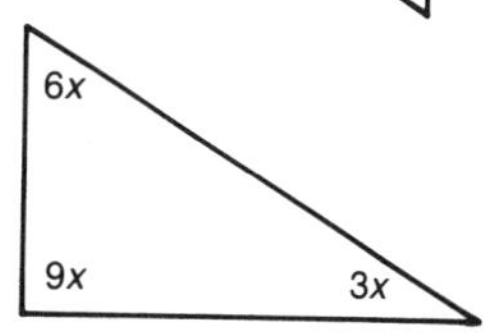


**61.**

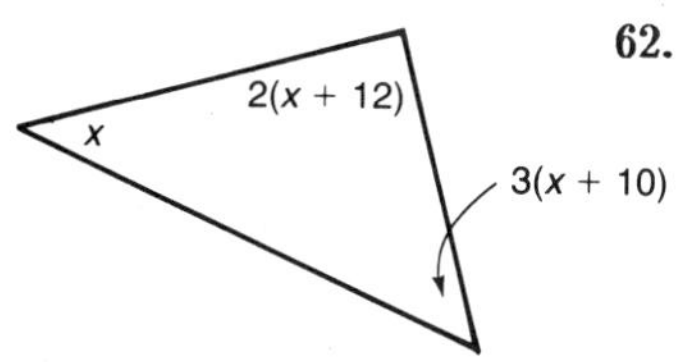


**62.**

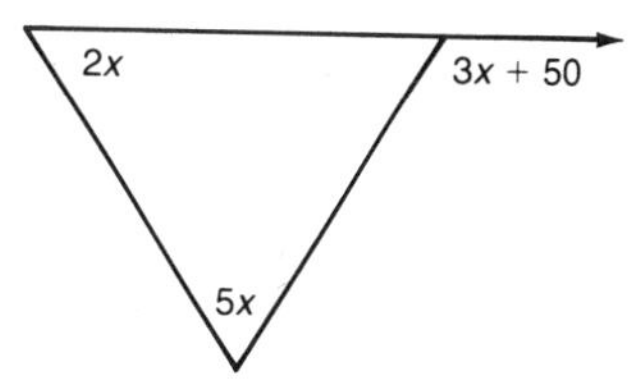


**63.**

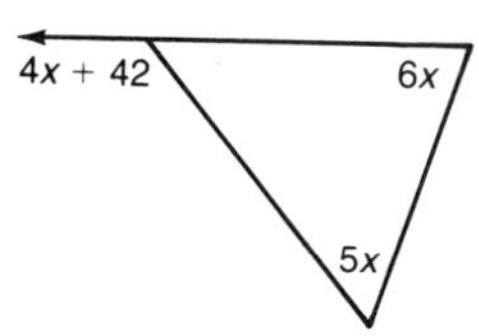


**64.**

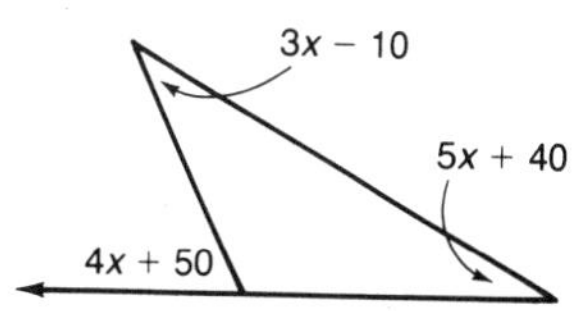


**65.** Find the measures of the angles of a triangle if it is known that the measures of the angles are $x$, $14 + 3x$, and $3(x + 25)$.

**66.** Find the measures of the angles of a triangle if it is known that the measures of the angles are $x$, $3x - 10$, and $3(55 - x)$.

# CHAPTER 2 MEASURING LENGTH

## 2.1 Precision and Accuracy

It is necessary to use measurement in geometry problems. Often, you will have to determine the length, width, height, area, or volume of a geometric figure. Measurement, however, is never exact, and you will need to decide how precise the measurement should be. For example, the measurement might be to the nearest inch, nearest foot, or even nearest mile. The **precision** of a measurement depends not only on the instrument used, but also on the purpose of your measurement. For example, if you are measuring the size of a room in order to lay carpet, the precision of your measurement might be different than if you are measuring the size of an airport hangar.

The **accuracy** refers to your answer. Suppose you measure with an instrument that measures to the nearest tenth of a unit. You find one measurement to be 4.6 and another to be 2.1. If, in the process of your work, you need to multiply these numbers, the result you obtain is

$$4.6 \times 2.1 = 9.66$$

This product is calculated to two decimal places, but it does not seem quite right that you obtain an answer that is more accurate (two decimal places) than the instrument you are using to do your measurement (one decimal place). In this book, we will require that the accuracy of your answers not exceed the precision of the measurement. This means that, after doing your calculations, you should round the final answer. The principle we will use is stated in the following box.

**ACCURACY OF THE ANSWERS IN THIS BOOK**

> All measurements are as precise as given in the text. If you are asked to make a measurement, the precision will be specified. Carry out all calculations *without rounding*. After obtaining a *final answer*, round this answer to be as accurate as the *least precise measurement*.

To avoid round-off error, you should round only once (at the end). This is particularly important now that calculators are available and calculations are often displayed to 8, 10, or 12 decimal places.

## 2.2 Distance Around

*Length* is actually a measure of the distance from one place to another.

**MEASURE OR LENGTH**

> To measure an object is to assign a number to its size. The number representing its linear dimensions as measured from end to end is called its **measure or length.**

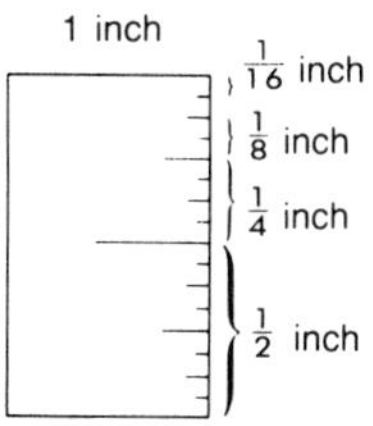


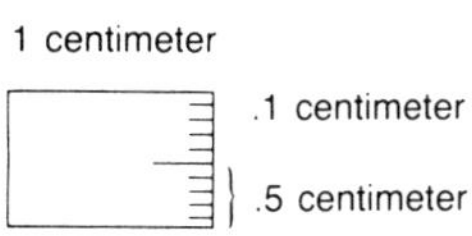


In order to measure length, you need a measuring device appropriate to the length you are measuring. In the United States, we use a system of measurement that was originally based on parts of the human body. In this system of measurement, length is measured in units such as inches (in.), feet (ft), yards (yd), and miles (mi). However, the rest of the world uses the **metric system.** The real advantage of the metric system is the ease of conversion from one unit of measurement to another. The centimeter (cm), meter (m), and kilometer (km) are commonly used metric units of length.

### Perimeter

The **perimeter** of a polygon is the distance around the polygon. It is the sum of the lengths of the sides of that polygon. Here are some formulas for finding the perimeters of the most common polygons.

**PERIMETER FORMULAS**

A **triangle** is a polygon with three sides.

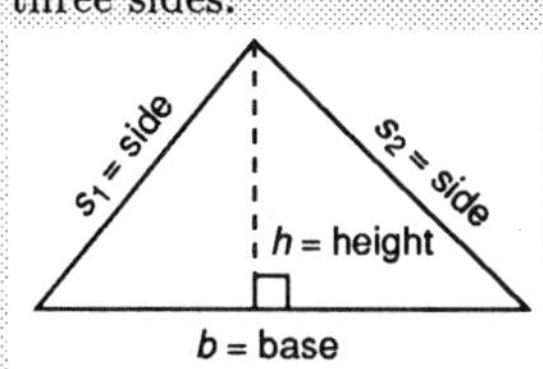


$P = s_1 + s_2 + b$

An **equilateral triangle** is a triangle with sides that are equal.

$P = 3s$

A **rectangle** is a quadrilateral with all right angles.

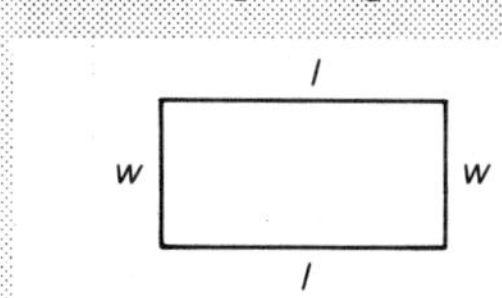


$P = 2l + 2w$

A **square** is a rectangle with sides that are equal.

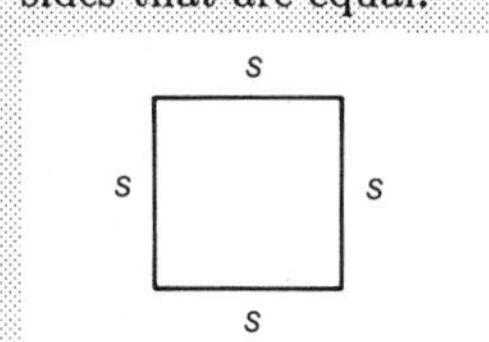


$P = 4s$

**EXAMPLE 1** Find the perimeter of each polygon.

**a.**

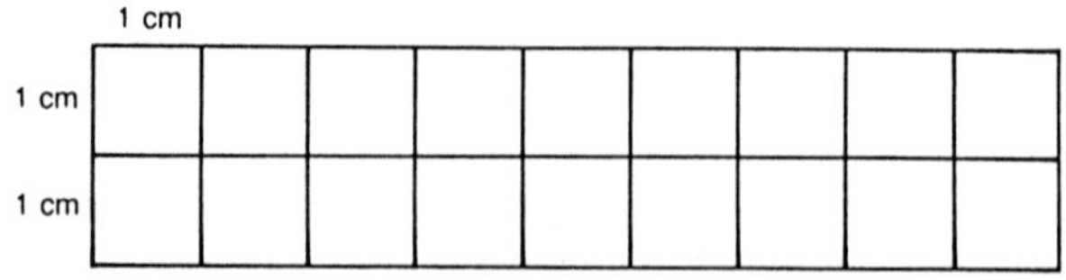


**b.**

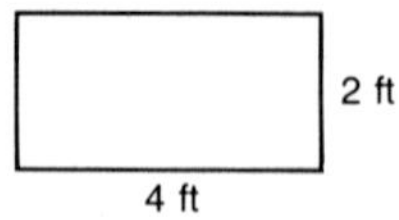


**c.**

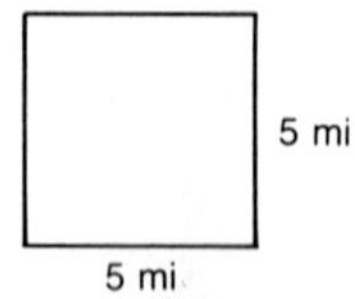


**d.**

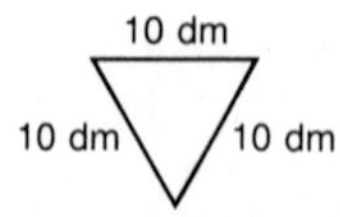


**e.**

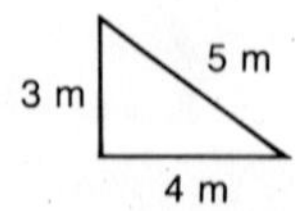


*Solution*

**a.** Rectangle is 2 cm by 9 cm,

$$P = 2l + 2w$$
$$= 2(9) + 2(2)$$
$$= 22$$

The perimeter is 22 cm.

**b.** Rectangle is 2 ft by 4 ft, so

$$P = 2l + 2w$$
$$= 2(4) + 2(2)$$
$$= 12$$

The perimeter is 12 ft.

**c.** Square, so

$$P = 4s$$
$$= 4(5)$$
$$= 20$$

The perimeter is 20 mi.

**d.** Equilateral triangle, so

$$P = 3s$$
$$= 3(10)$$
$$= 30$$

The perimeter is 30 dm.

**e.** Triangle, so

$$P = 3 + 4 + 5 = 12$$

The perimeter is 12 m. □

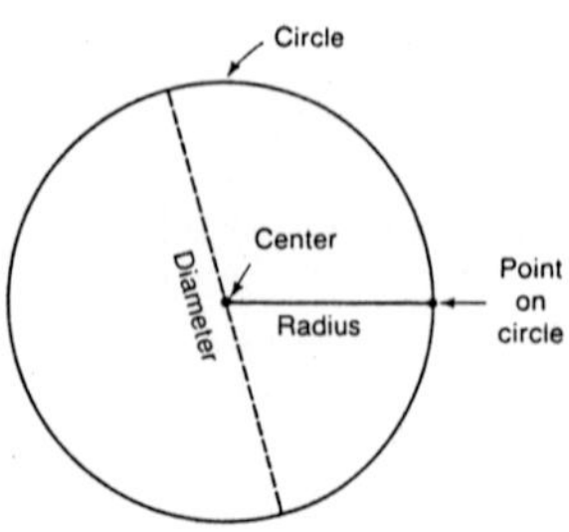


Figure 17

## Circumference

A **circle** is not a polygon, but you will need to find the distance around a circle or part of a circle. Figure 17 shows some of the terminology associated with a circle. The distance around the circle is called its **circumference** and is found by using the following formula.

**CIRCUMFERENCE**

The circumference C of a circle with radius $r$ and diameter $d$ is

$$C = d\pi \quad \text{or} \quad C = 2\pi r$$

**EXAMPLE 2** Find the circumference of each circle (or the distance around each figure).

**a.**

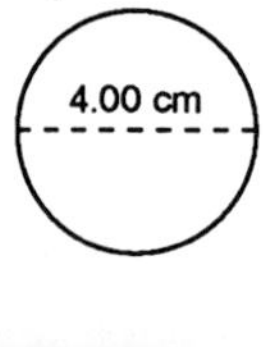


**b.**

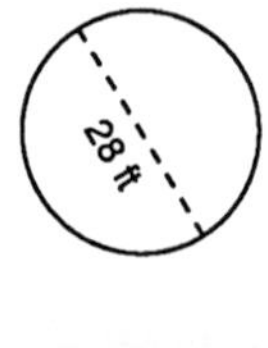


**c.**

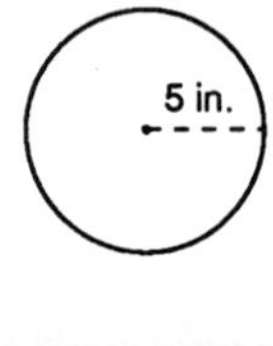

d. 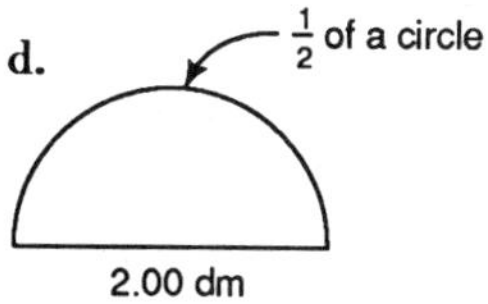


e. 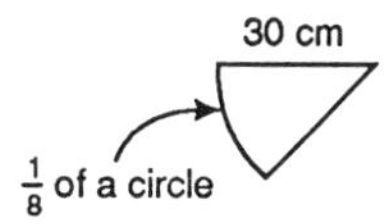


*Solution*

a. $C = 4\pi$. This is called the **exact** answer. A decimal approximation is $C \approx 4(3.1416) \approx 12.56$. The circumference is $4\pi$ cm or about 12.56 cm.

b. $C = 28\pi$; this is about 88 ft.

c. $C = 2\pi(5) = 10\pi$; this is about 30 in.

d. This is half of a circle (called a **semicircle**); thus, the curved part is half of the circumference ($C = 2\pi \approx 6.28$), or about 3.14, which is added to the diameter:

$$\tfrac{1}{2}(2\pi) + 2 \approx 5.1415927 \quad \textit{by calculator}$$

The distance around the figure is 5.14 dm (answer rounded).

e. This is one-eighth of a circle. The curved part is one-eighth of the circumference ($C = 2\pi(30) \approx 188.4$), or about 23.6. Thus, the distance around the figure (do not forget to add the two sides of the figure) is

$$\tfrac{1}{8}[2\pi(30)] + 2(30) \approx 83.561945 \quad \textit{by calculator}$$

The distance around the figure is 84 cm (answer rounded). □

**Irregular shapes**

You may be asked to find the perimeter (or distance around) an object that is not a single, regular shape. If the irregular object has line segments as sides, you can find the perimeter by adding together the lengths of all the sides. Some irregular-shaped objects have sides that are not line segments, but parts of a circle.

**EXAMPLE 3** Find the distance around the figure shown at right.

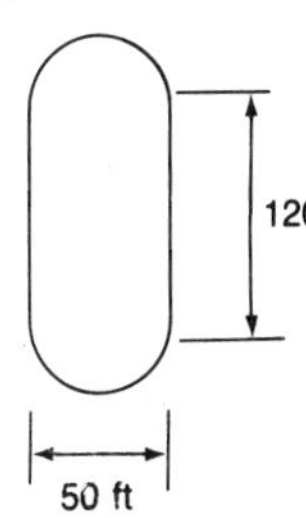


*Solution*

$$\text{DISTANCE AROUND} = \pi(50) + 120 + 120 = 240 + 50\pi \approx 397.07963$$

The distance around is 397 ft (answer rounded). □

**EXAMPLE 4** Find the perimeter of the figure shown at right.

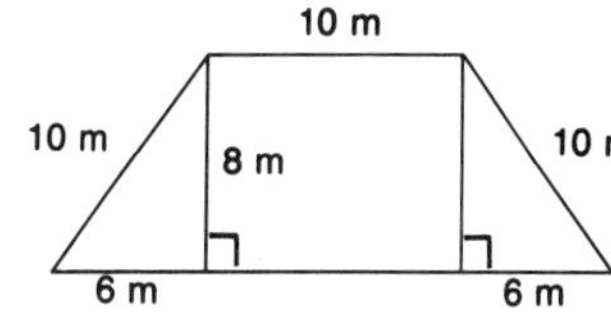


*Solution* The perimeter of an irregularly shaped figure can be found if you know the lengths of the sides.

$P = 10 + 10 + 10 + 6 + 10 + 6 = 52$

The perimeter is 52 m. □

**EXAMPLE 5** Use the figure given at right to find the circumference of the circle if you know that the perimeter of the square is 20.

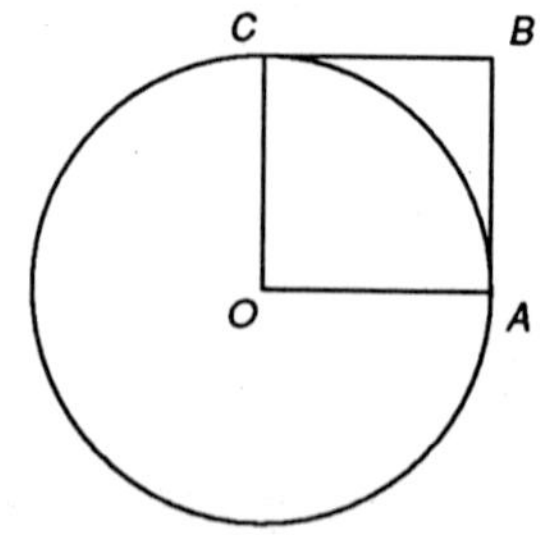


*Solution* The perimeter of the square is 20, so

$$\begin{aligned} P &= 4s \\ 20 &= 4s \\ 5 &= s \end{aligned}$$

Notice that a side of the square is a radius of the circle so

$$\begin{aligned} C &= 2\pi r \\ &= 2\pi(5) \\ &= 10\pi \end{aligned}$$

This is approximately 31 inches. □

## 2.3 Solving Geometry Problems

When solving geometry problems, you should have a systematic approach. The following steps will help:

1. Draw a picture of the geometric figure described in the problem.
2. Label the measurements on the picture.
3. Determine the unknown — what it is that you are asked to find.
4. Identify a formula that will allow you to solve for the unknown.
5. Substitute the known values into the formula.
6. Solve the equation for the unknown value.
7. Check your answer to make sure that it is reasonable and that it has the correct units. Remember that perimeter and circumference are measured in linear units, area in square units, and volume in cubic units.

**EXAMPLE 6** Suppose you have enough material for 70 ft of fence and want to build a rectangular pen 14 ft wide. What is the length of this pen?

*Solution* Draw and label a figure as shown.

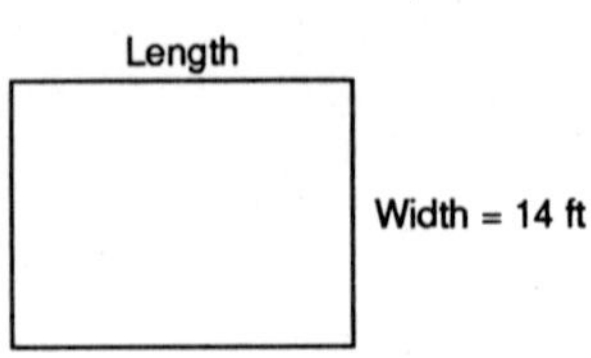


$$\begin{aligned} \text{PERIMETER} &= 2(\text{LENGTH}) + 2(\text{WIDTH}) \\ 70 &= 2(\text{LENGTH}) + 2(14) \end{aligned}$$

Let $l$ = LENGTH

$$\begin{aligned} 70 &= 2l + 28 \\ 42 &= 2l \\ 21 &= l \end{aligned}$$

The pen will be 21 ft long. □

## 2.4 PROBLEM SET 2

**A** *Find the perimeter or circumference of the figures given in Problems 1–15 by using the appropriate formula. Round approximate answers to two decimal places.*

**1.**
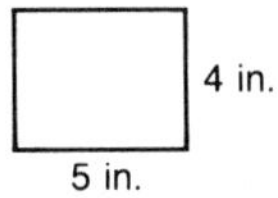


**2.**
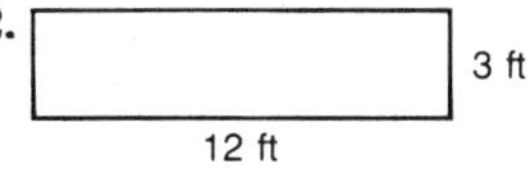


**3.**
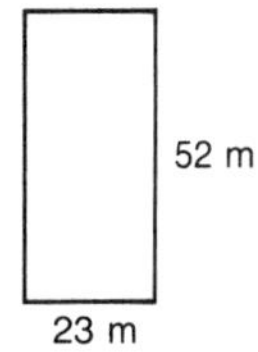


**4.**
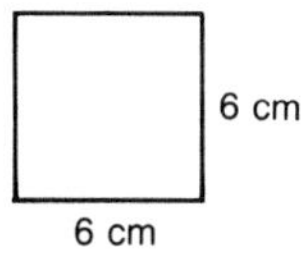


**5.**
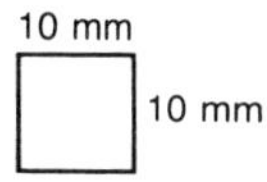


**6.**
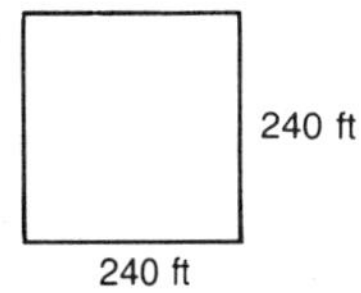


**7.**
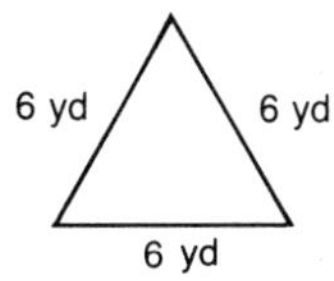


**8.**
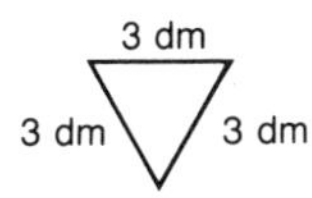


**9.**
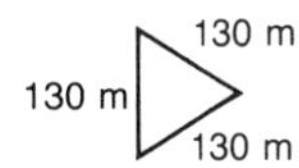


**10.**
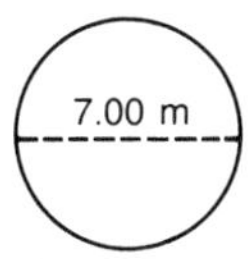


**11.**
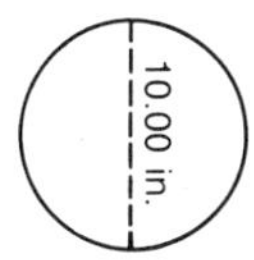


**12.**

**13.**
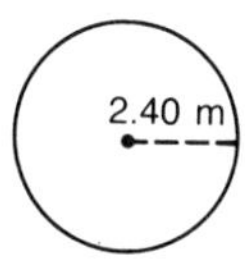


**14.**
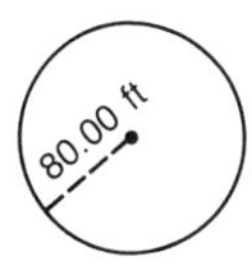


**15.**
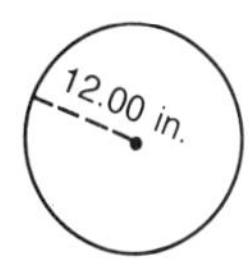


*Find the distance around the figures in Problems 16–27. Round approximate answers to one decimal place.*

**16.**
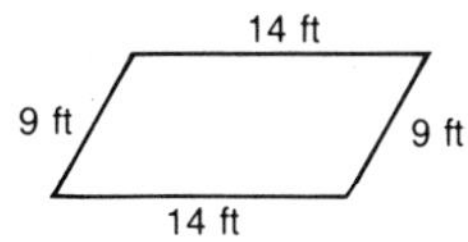


**17.**
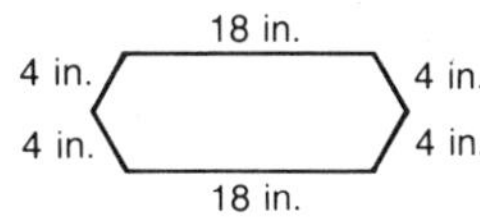


**18.**
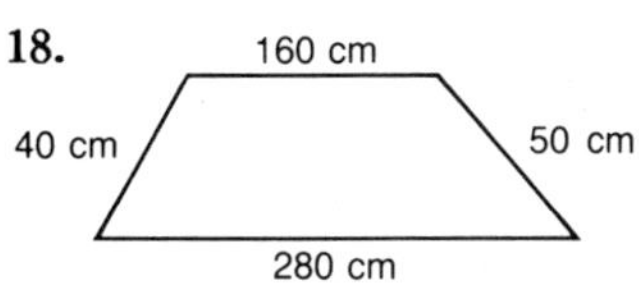


**19.**
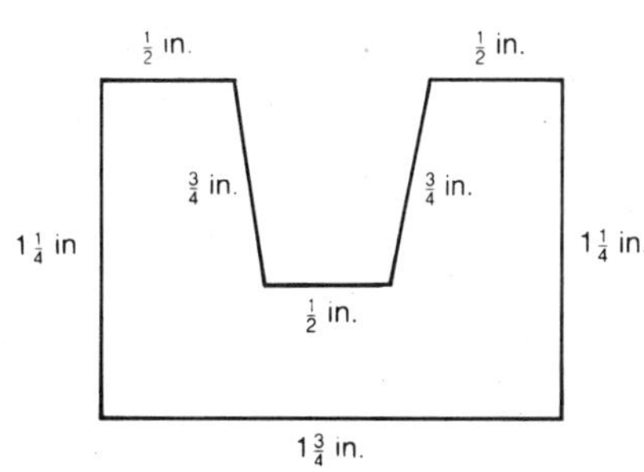

**20.**

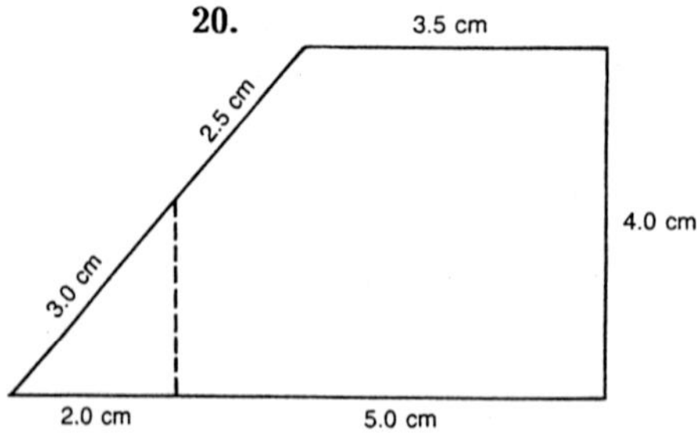


**21.**

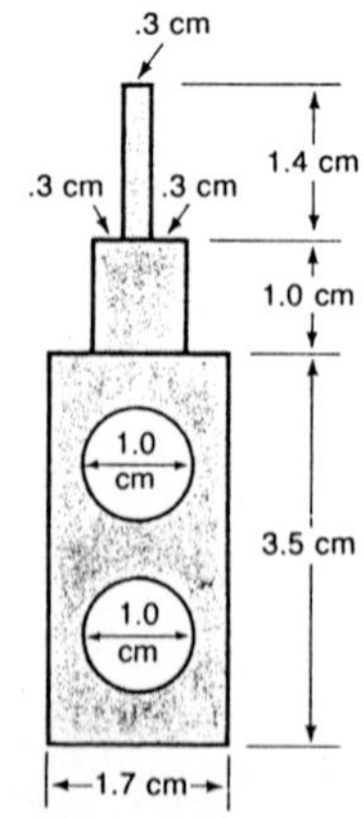


**22.**

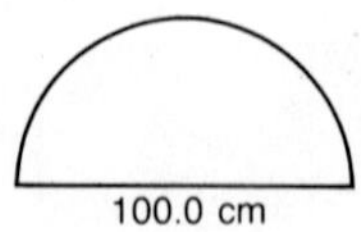


**23.**

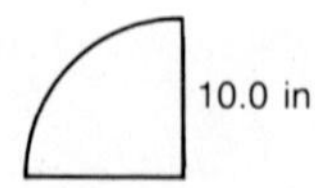


**24.**

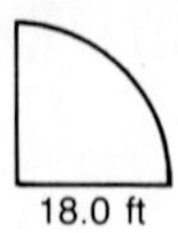


**25.**

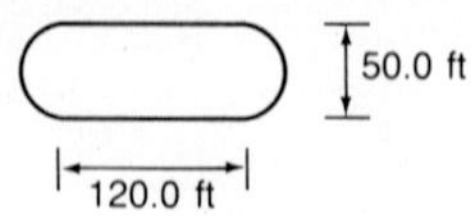


**26.**

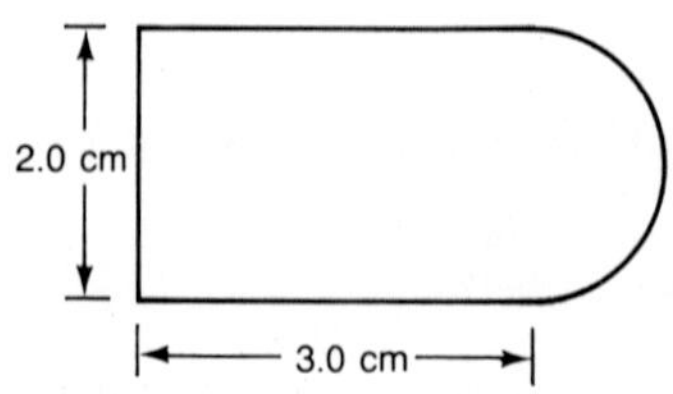


**27.**

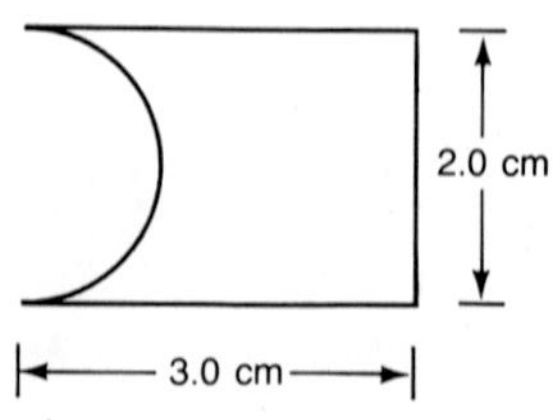


*Find x in Problems 28–31.*

**28.** Perimeter is 92 ft.

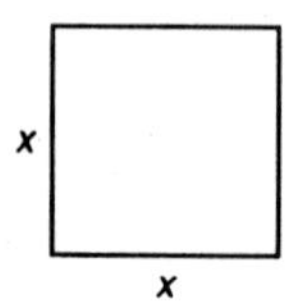


**29.** Perimeter is 102 in.

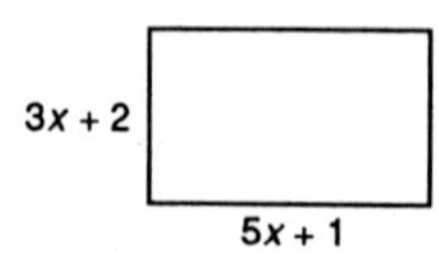


**30.** Circumference is 392.7 cm.

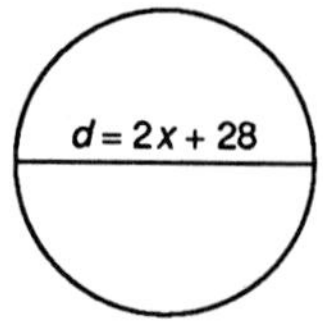


**31.** Perimeter is 219.6 mm.

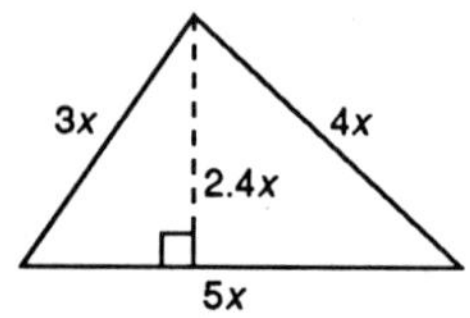

**B**

**32.** What is the width of a rectangular lot that has a perimeter of 410 ft and a length of 140 ft?

**33.** What is the length of a rectangular lot that has a perimeter of 750 m and a width of 75 m?

**34.** The perimeter of $\triangle ABC$ is 117 in. Find the lengths of the sides.

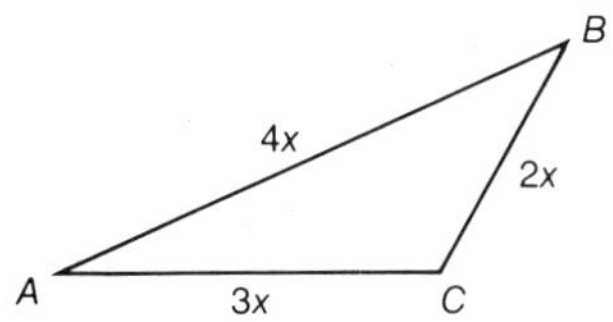


**35.** The perimeter of the pentagon is 280 cm. Find the lengths of the sides.

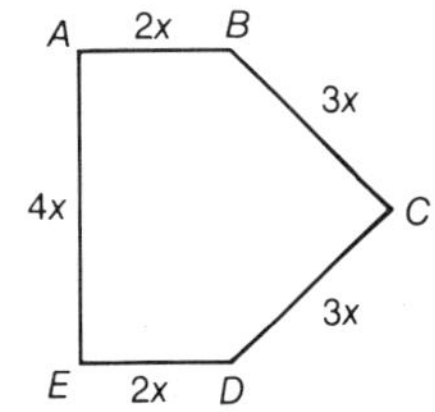


**36.** Use the figure below to find the area of the quartercircle if the perimeter is $4 + \pi$.

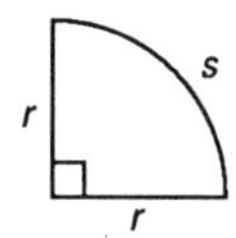


**37.** A regular hexagon is formed from 6 equilateral triangles, as shown below. If each triangle has a perimeter of 4, what is the perimeter of the hexagon?

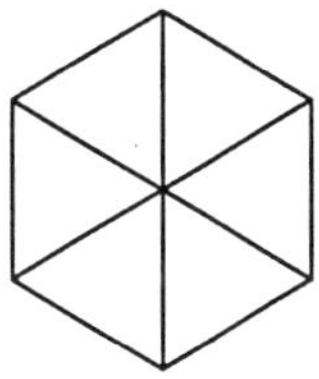

**38.** Find the dimensions of an equilateral triangle that has a perimeter measuring 198 dm.

**39.** Find the dimensions of a rectangle with a perimeter of 54 cm if the length is 5 less than three times the width.

# CHAPTER 3 MEASURING AREA

## 3.1 Square Units

**Area** is a measurement of how much space is inside a two-dimensional object. For example, suppose you want to carpet your living room. The price of carpet is quoted in terms of a price *per square yard.* A square yard is a measure of area. The unit of measurement used for area depends on the unit you choose for measuring the length of the sides. For example, if you choose the yard as the unit of length for the dimensions of a room when ordering carpet, the fundamental unit of area is then a square yard. A square yard is a measure of the amount of space inside a square that is 1 yard long on each side. The measurement 1 square yard is abbreviated 1 sq yd or 1 $yd^2$. We will more commonly use square inches or square centimeters, as shown in Figure 18.

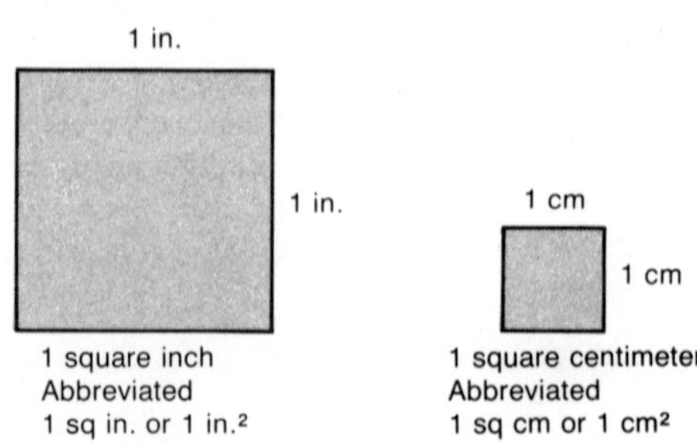


Figure 18

## 3.2 Rectangles and Squares

### Area by counting

**EXAMPLE 1** What is the area of the shaded region?

*Solution*

**a.**

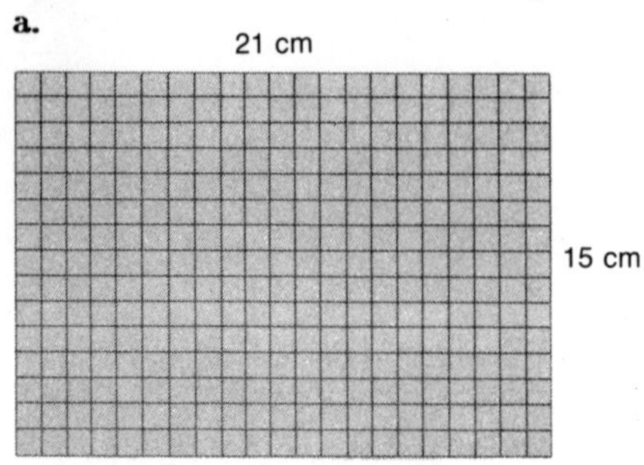


You can count the number of square cm in the shaded region; there are 315 squares. Also notice:

*Across* *Down*

21 cm × 15 cm

$= 21 \times 15 \times \text{cm} \times \text{cm}$

$= 315\ \text{cm}^2$

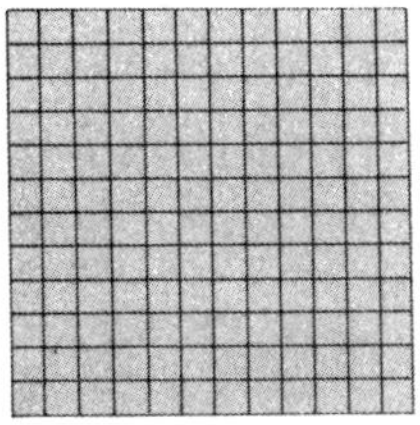


b. The shaded region is a **square foot.** You can count 144 square inches inside the region. Also notice:

*Across* *Down*
12 in. × 12 in. = 144 in.$^2$ □

**Area by formula**

As you can see from Example 1, the area of a rectangle or square region is the product of the distance across (length) and the distance down (width).

AREA OF RECTANGLES AND SQUARES

The formula for finding the area of a rectangle is:

$$A = lw$$

The formula for finding the area of a square is:

$$A = s^2$$

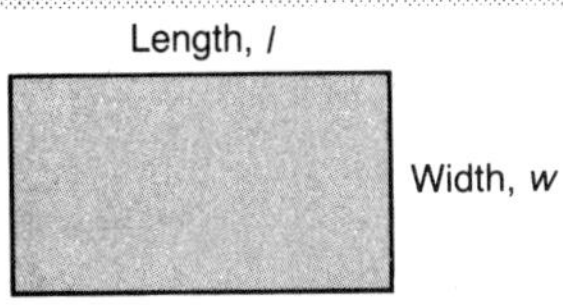


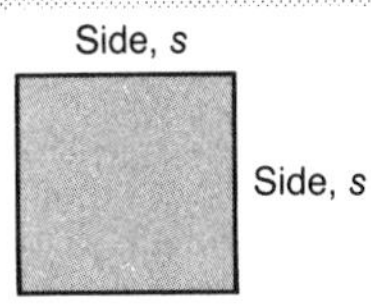


EXAMPLE 2

How many square feet are there in 1 square yard?

*Solution*

Since 1 yd = 3 ft,

$1 \text{ yd}^2 = (3 \text{ ft})^2 = 9 \text{ ft}^2$

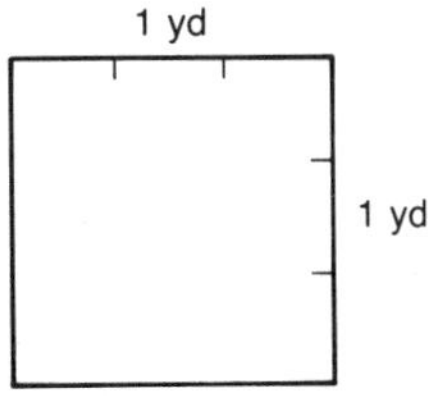


□

**Converting units**

Sometimes we measure area using one unit of measurement and then want to convert the result to another unit of measurement. Consider the following example.

EXAMPLE 3

Suppose your living room is 12 ft by 15 ft, and you want to know how many square yards of carpet you need to cover this area.

*Solution*

METHOD I: Calculate; then convert.

$$\begin{aligned} A &= lw \\ &= 12 \text{ ft} \times 15 \text{ ft} \\ &= 180 \text{ ft}^2 \\ &= 180 \times 1 \text{ ft}^2 \\ &= 180 \times (\tfrac{1}{9} \text{ yd}^2) && \textit{1 yd = 3 ft, so 1 yd}^2 = \textit{9 ft}^2\textit{, so} \\ &= 20 \text{ yd}^2 && \tfrac{1}{9}\ \textit{yd}^2 = \textit{1 ft}^2 \end{aligned}$$

METHOD II: Convert; then calculate.

12 ft = 4 yd and 15 ft = 5 yd, so

$$A = lw$$
$$= 4 \text{ yd} \times 5 \text{ yd}$$
$$= 20 \text{ yd}^2$$

□

**Acres**

If the area is large, as with property, a larger unit is needed. This is called an **acre.**

**ACRE**

An **acre** is 43,560 $\text{ft}^2$.

To change from $\text{ft}^2$ to acres, divide by 43,560.

When working with acres, you usually need to have a calculator handy in order to change from square feet to acres.

**EXAMPLE 4** How many acres are there in a rectangular piece of property measuring 363 ft by 180 ft?

*Solution*

$$A = lw$$
$$= 363 \text{ ft} \times 180 \text{ ft}$$
$$= 65{,}340 \text{ ft}^2$$

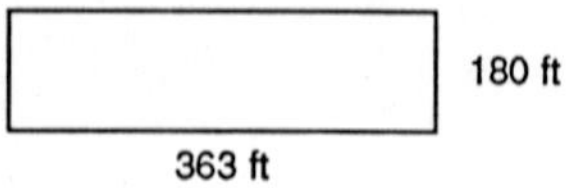


To change from $\text{ft}^2$ to acres, divide by 43,560:

$$A = 65{,}340 \text{ ft}^2$$
$$= (65{,}340 \div 43{,}560) \text{ acres}$$
$$= 1.5 \text{ acres}$$

□

### 3.3 Parallelograms

Remember, a parallelogram is a quadrilateral with two pairs of parallel sides, as shown in Figure 19.

Figure 19

To find the area of a parallelogram, you can count the squares necessary to fill the region, rearrange part of the given parallelogram so that it forms a rectangle, or use the formula given in the following box.

**AREA OF PARALLELOGRAM**

The formula for finding the area of a parallelogram is

$$A = bh$$

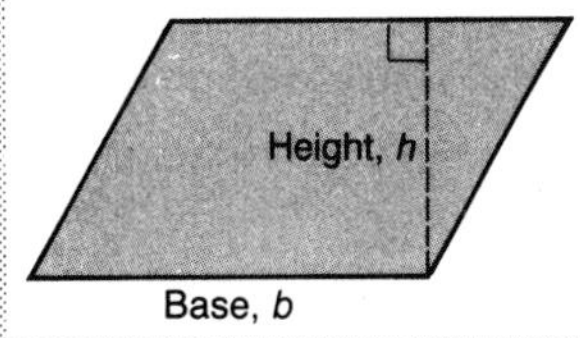


**EXAMPLE 5**

Find the area of each shaded region.

**a.**

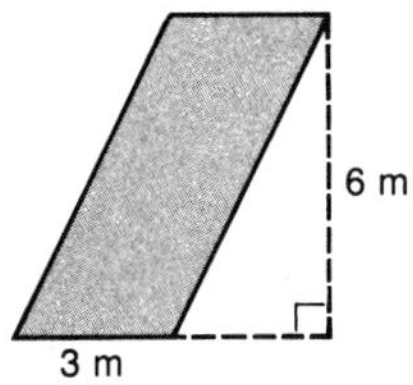


**b.**

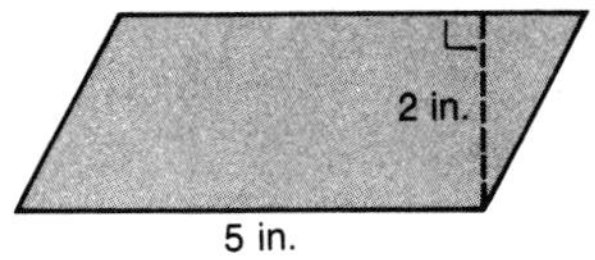


*Solution*

**a.** $A = 3 \text{ m} \times 6 \text{ m}$
$= 18 \text{ m}^2$

**b.** $A = 2 \text{ in.} \times 5 \text{ in.}$
$= 10 \text{ in.}^2$ □

### 3.4 Triangles

You can find the area of a triangle by filling in square units or by rearranging the parts, but the easiest way is to notice that *every* triangle has an area that is exactly half that of the corresponding parallelogram. We can therefore state the formula for the area of a triangle as in the following box.

**AREA OF TRIANGLE**

The formula for finding the area of a triangle is

$$A = \tfrac{1}{2}bh$$

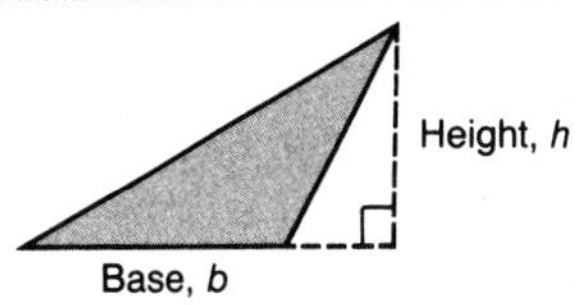


**EXAMPLE 6**

Find the area of each shaded region.

**a.**

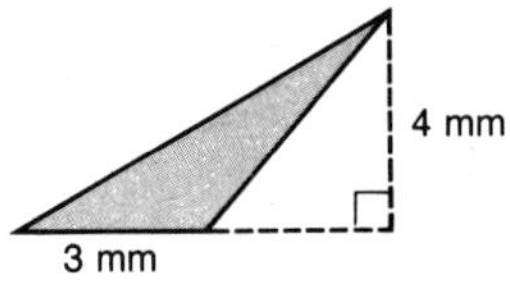


**b.**

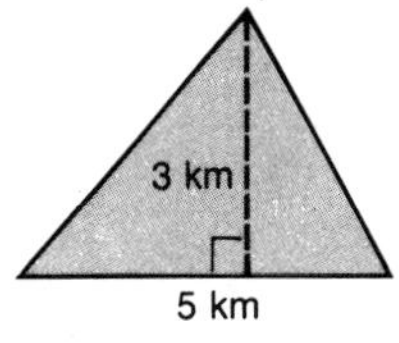


*Solution*

**a.** $A = \tfrac{1}{2}bh$
$= \tfrac{1}{2}(4 \text{ mm})(3 \text{ mm})$
$= 6 \text{ mm}^2$

**b.** $A = \tfrac{1}{2}bh$
$= \tfrac{1}{2}(3 \text{ km})(5 \text{ km})$
$= \tfrac{15}{2} \text{ km}^2 \text{ or } 7\tfrac{1}{2} \text{ km}^2$

□

**EXAMPLE 7**

What is the area of the following figure, if $AC = 4$, $OD = 2$, $OB = 6$, and $AC$ is perpendicular to $BD$?

*Solution* First, label the given distances:

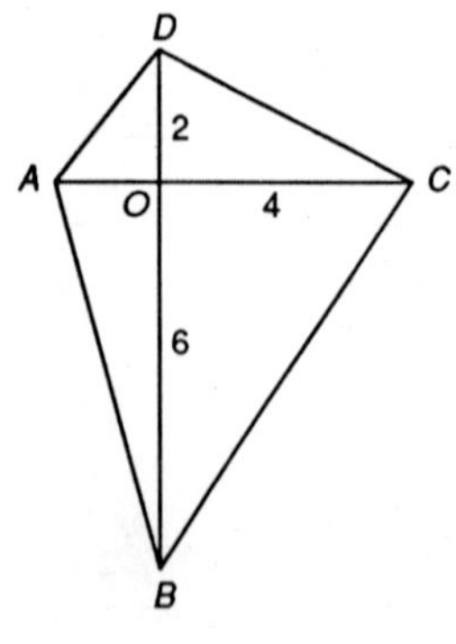


The area is the sum of the area of two triangles:

Area of $\triangle ADC = \frac{1}{2}bh = \frac{1}{2}(4)(2) = 4$

Area of $\triangle ABC = \frac{1}{2}bh = \frac{1}{2}(4)(6) = 12$

Total area $= 4 + 12 = 16$

□

### 3.5 Trapezoids

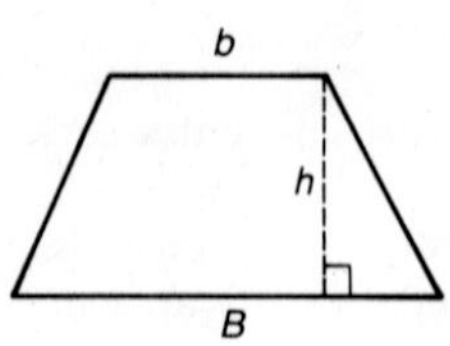


Figure 20

We can also find the area of a trapezoid by finding the area of triangles. A **trapezoid** is a quadrilateral with two sides parallel. These sides are called the *bases*, and the perpendicular distance between the bases is the height, as shown in Figure 20. The area formula can be found as the sum of the areas of triangle I and triangle II, as shown in Figure 21.

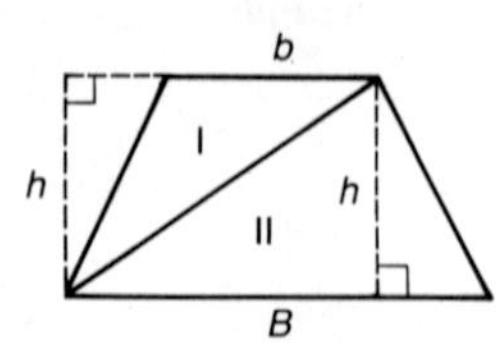


Area of triangle I: $\frac{1}{2}bh$

Area of triangle II: $\frac{1}{2}Bh$

Total area: $\frac{1}{2}bh + \frac{1}{2}Bh$

Figure 21

If we use the distributive property, we can obtain the area formula for a trapezoid, as shown in the following box.

**AREA OF TRAPEZOID**

The formula for finding the area of a trapezoid is:

$$A = \frac{1}{2}h(b + B)$$

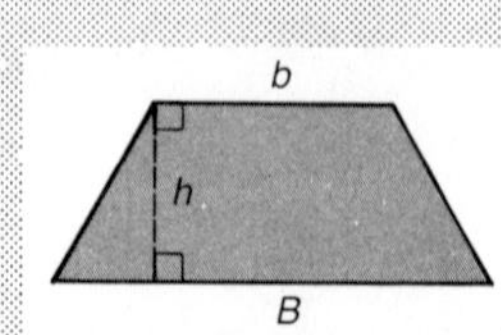


EXAMPLE 8 Find the area of each shaded region.

*Solution* **a.**

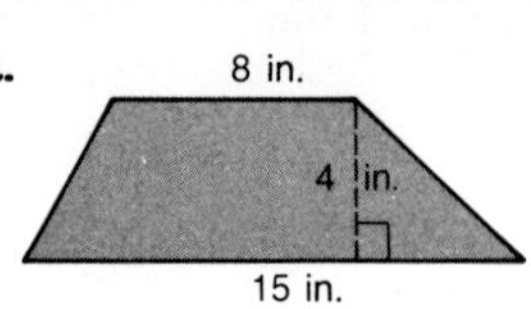


$h = 4$ in.; $b = 8$ in.; $B = 15$ in.

$$\begin{aligned} A &= \tfrac{1}{2}(4)(8 + 15) \\ &= 2(23) \\ &= 46 \end{aligned}$$

The area is 46 in.$^2$.

b.

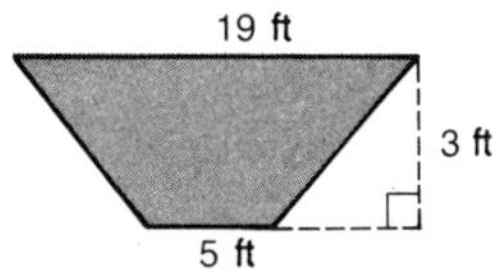


$h = 3$ ft; $b = 19$ ft; $B = 5$ ft

$$\begin{aligned} A &= \tfrac{1}{2}(3)(19 + 5) \\ &= \tfrac{3}{2}(24) \\ &= 36 \end{aligned}$$

The area is 36 $\text{ft}^2$. □

### 3.6 Circles

To find the area of a circle, you must use a formula.

**AREA OF CIRCLE**

The formula for finding the area of a circle is:

$$A = \pi r^2$$

Note: If you are given the diameter, remember that $r = \frac{1}{2}d$.

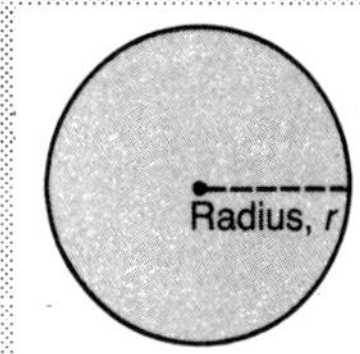


**EXAMPLE 9**

Find the area of each shaded region to the nearest tenth unit.

a.

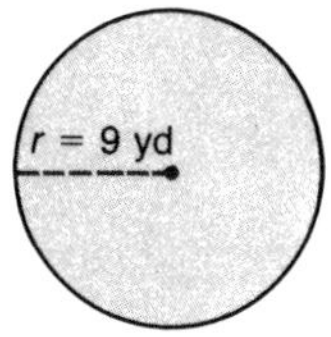


b.

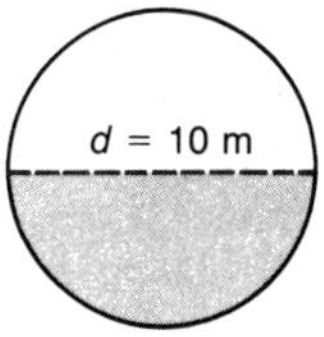


*Solution*

a. $$\begin{aligned} A &= \pi(9)^2 \\ &= 81\pi \\ &\approx 254.469 \end{aligned}$$

b. $d = 10$ m, so $r = 5$ m

$$\begin{aligned} A &= \tfrac{1}{2}\pi(5)^2 \\ &= \tfrac{25}{2}\pi \\ &\approx 39.269908 \end{aligned}$$

These approximations were found by using a calculator. Now round these answers to the requested accuracy.

a. The area is 254.5 $\text{yd}^2$. b. The area is 39.3 $\text{m}^2$. □

**EXAMPLE 10**

If the area of circle $X$ is twice the area of circle $Y$, and if the radius of circle $X$ is 4, what is the radius of circle $Y$?

*Solution*

Let $x$ be the radius of circle $X$ and $y$ be the radius of circle $Y$. Since the area formula is $A = \pi r^2$, we have

$$\begin{aligned} \pi x^2 &= 2\pi y^2 \\ \pi(4)^2 &= 2\pi y^2 \\ 16\pi &= 2\pi y^2 \\ 8 &= y^2 \\ 2\sqrt{2} &= y \end{aligned}$$

*Square root property; disregard the negative value, since $y$ is a distance.*

The radius of circle $Y$ is $2\sqrt{2}$. □

**EXAMPLE 11**

Find the area of the shaded portion of the figure shown at right, correct to two decimal places.

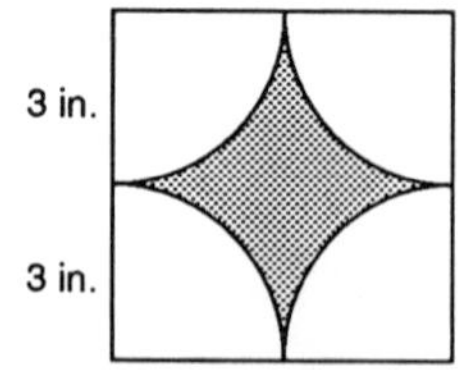


*Solution*

This figure can be rearranged as a circle of radius 3 inside the square with sides of length 6. The area of the shaded portion is the area of the rectangle minus the area of the circle:

$$A = 6^2 - \pi(3)^2$$
$$= 36 - 9\pi$$
$$\approx 7.7256661$$

*Use a calculator to find this number.*

The exact area is $36 - 9\pi$, which is approximately 7.73 in.$^2$. □

### 3.7 Surface Area

Suppose you wish to paint a small box whose edges are each 3 ft, and you need to know how much paint to buy. In order to do this, you need to find the sum of the areas of all the faces; this is called the **surface area.** To find the surface area, add the area of each face.

**EXAMPLE 12**

Find the amount of paint needed for a cube with edges 3 ft.

*Solution*

A cube has 6 faces of equal area. Each face is 9 ft$^2$, so

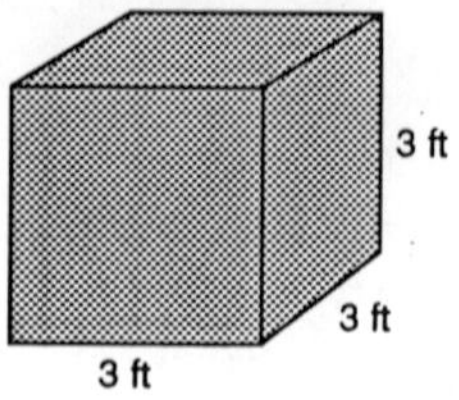


$S = 6(9 \text{ ft}^2) = 54 \text{ ft}^2$

You will need enough paint to cover 54 ft$^2$. □

**EXAMPLE 13**

Find the outside surface area.

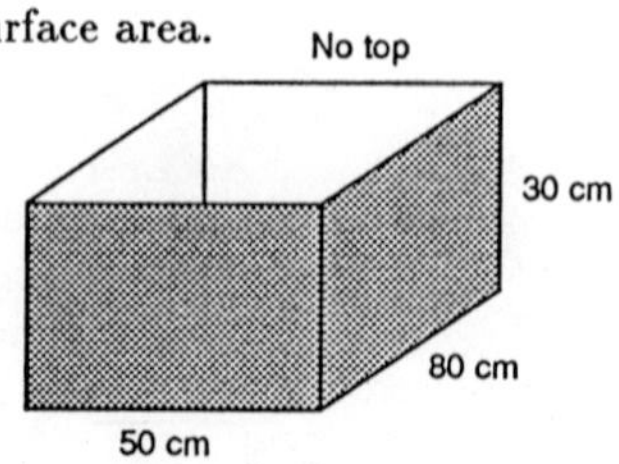


*Solution*

Notice that there is no top on the box.

| | |
|---|---|
| Bottom: | 80(50) = 4,000 |
| Front: | 50(30) = 1,500 |
| Back: | 50(30) = 1,500 |
| Side: | 80(30) = 2,400 |
| Side: | 80(30) = 2,400 |
| Total: | 11,800 cm |

□

**EXAMPLE 14**

Find the outside surface area.

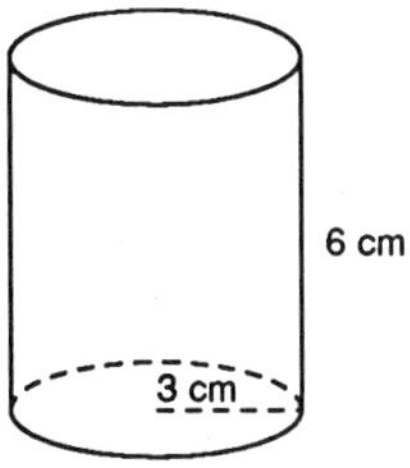


*Solution*

Notice that the can has a bottom and a top. To find the surface area, find the area of a circle (top and bottom), and think of the sides of the can as being "rolled out." The length of the resulting rectangle is the circumference of the can, and the width of the rectangle is the height of the can.

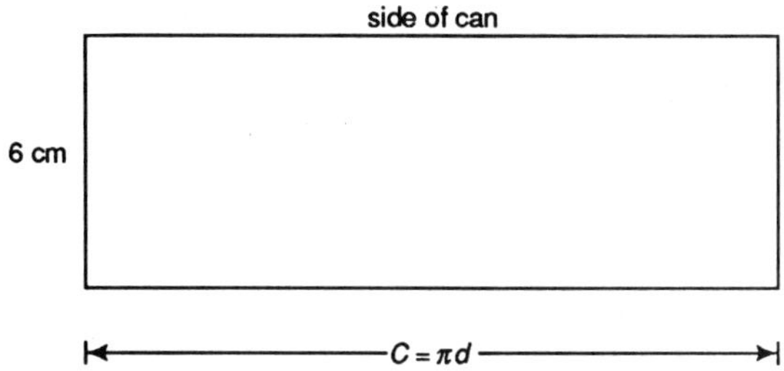


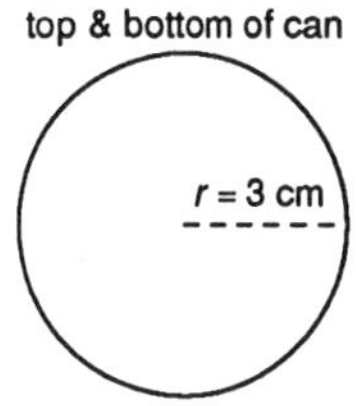


Side: $A = lw = (\pi d)w = \pi(6)(6) = 36\pi$
Top and Bottom: $A = \pi r^2 = \pi(3)^2 = 9\pi$ each

Surface area: $36\pi + 9\pi + 9\pi = 54\pi \approx 169.646$

The surface area is about 170 $\text{cm}^2$. □

**EXAMPLE 15**

A rectangular garden has a perimeter of 48 yards. The width of the garden is 6 yards less than its length. What is the area of the garden in square yards?

*Solution*

Draw a picture, and label the appropriate parts:

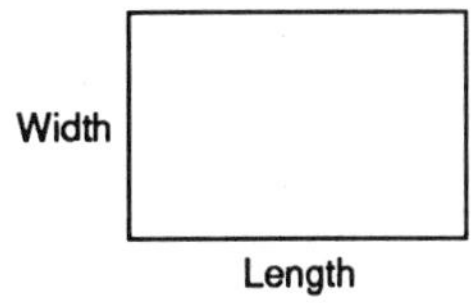


$$\begin{aligned} P &= 2(\text{LENGTH}) + 2(\text{WIDTH}) \\ &= 2(\text{LENGTH}) + 2(\text{LENGTH} - 6) && \textit{Substitute, since } \text{WIDTH} = \text{LENGTH} - 6 \\ &= 2l + 2(l - 6) && \text{Let } l = \text{LENGTH} \\ &= 4l - 12 \end{aligned}$$

Thus,

$$\begin{aligned} 4l - 12 &= 48 && \textit{Since the perimeter is 48 yards (given)} \\ 4l &= 60 \\ l &= 15 \end{aligned}$$

The width is $15 - 6 = 9$, so the area is

$$A = lw = 15(9) = 135$$

The area is 135 $\text{yd}^2$. □

## 3.8 PROBLEM SET 3

**A** *Find the area of each shaded region in Problems 1–22. (Assume that given measurements are exact, and round approximate answers to the nearest tenth of a square unit.)*

**1.**

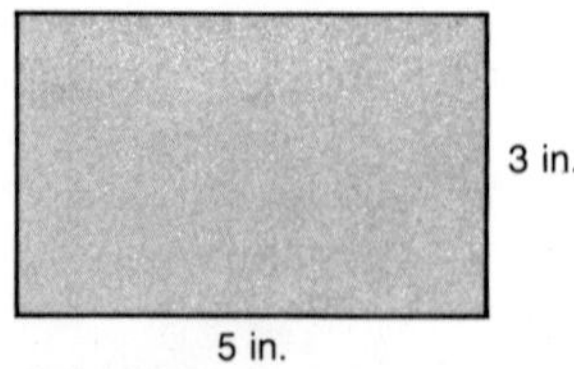


**2.**

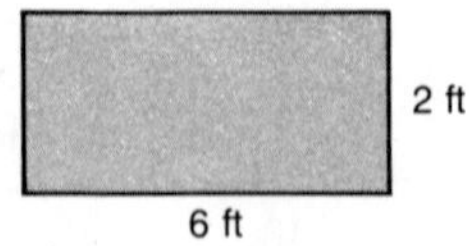


**3.**

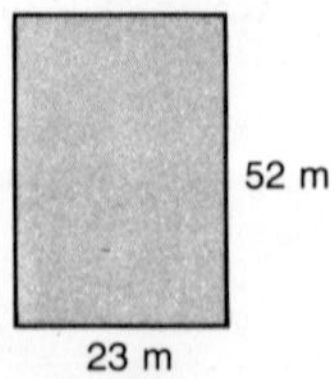


**4.**

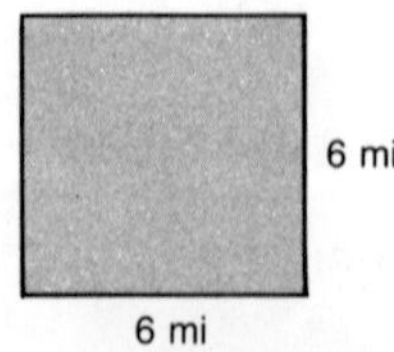


**5.**

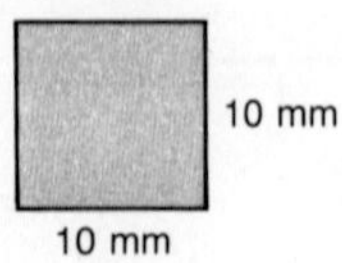


**6.**

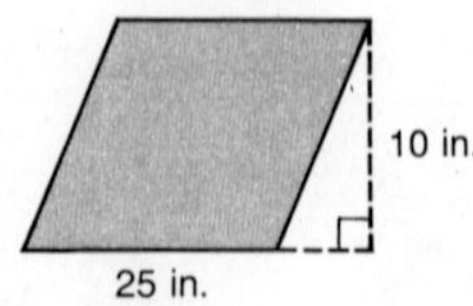


**7.**

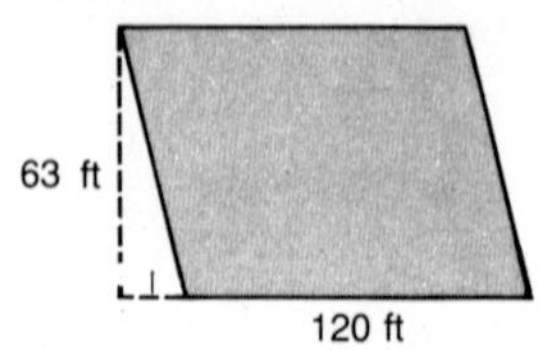


**8.**

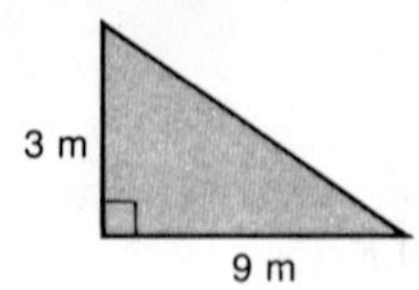


**9.**

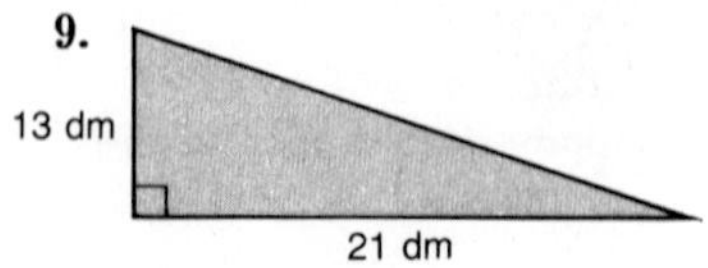


**10.**

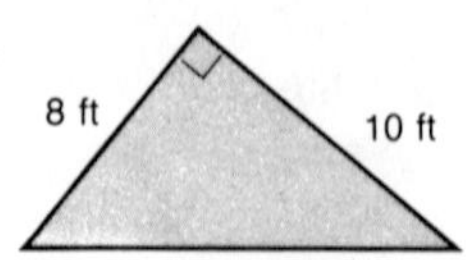


**11.**

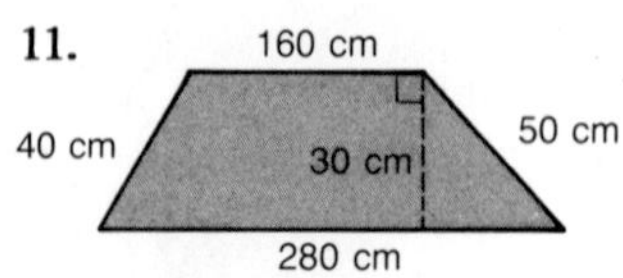


**12.**

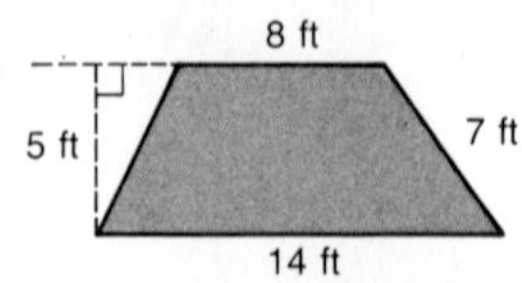


**13.**

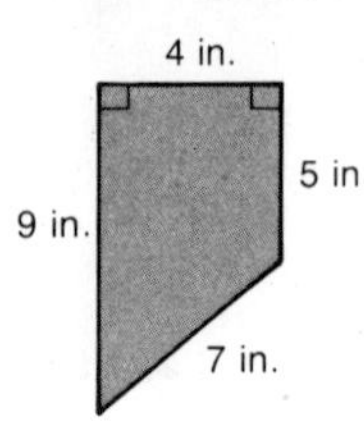


**14.**

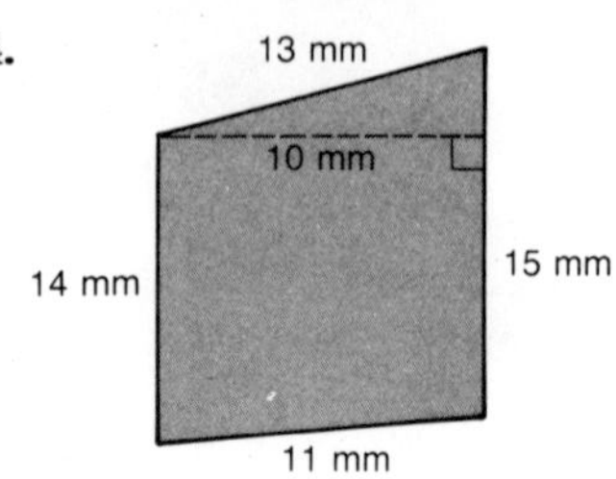

**15.**

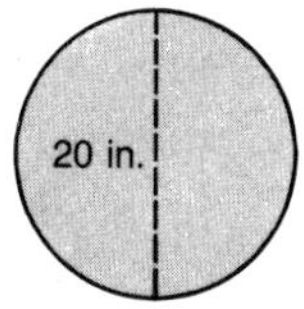


**16.**

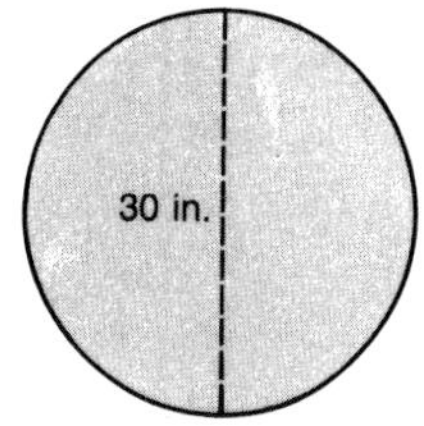


**17.**

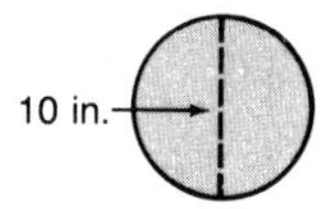


**18.**

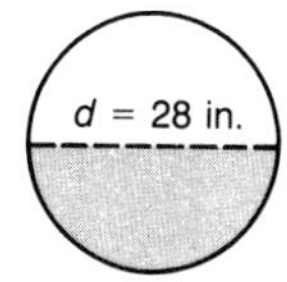


**19.**

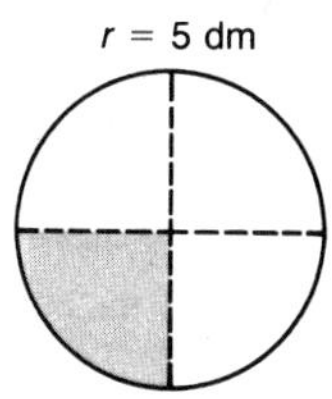


**20.**

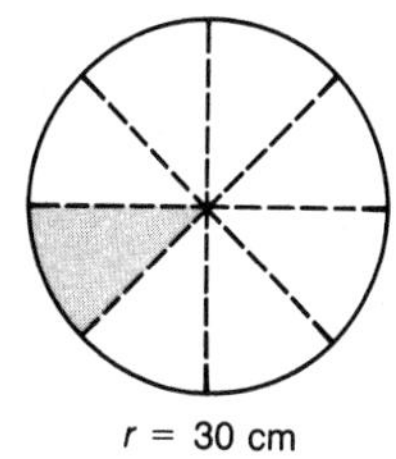


**21.**

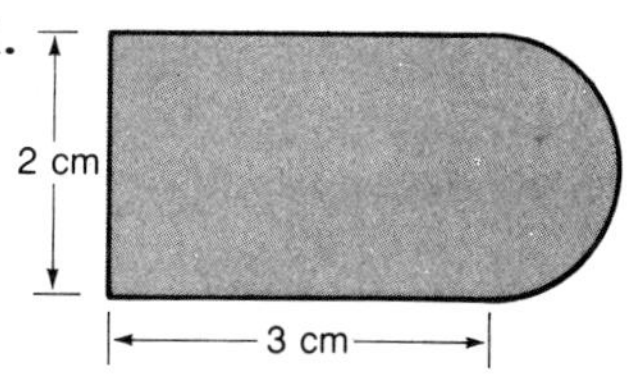


**22.**

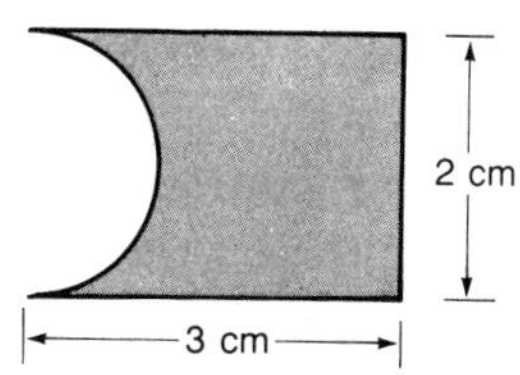


## B

*Find the outside surface areas in Problems 23–30. (Round your answer to the nearest square unit.)*

**23.**

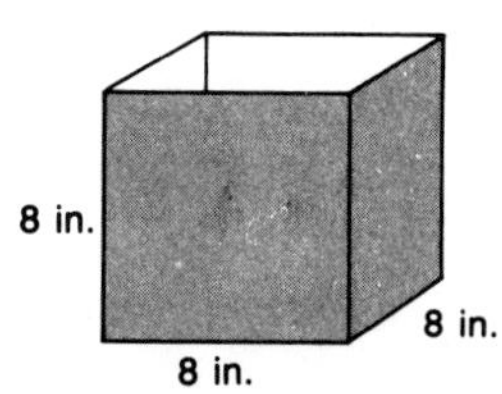


**24.**

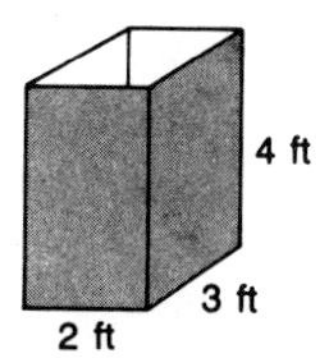


**25.**

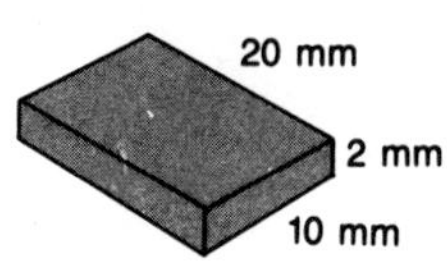


**26.**

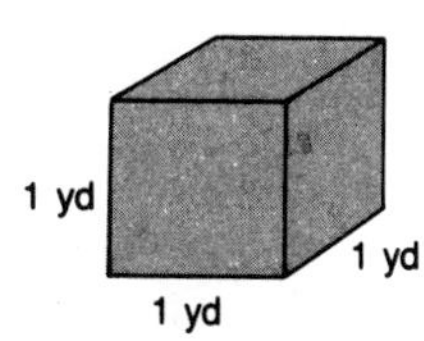


**27.**

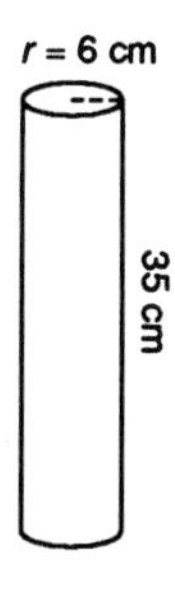


**28.**

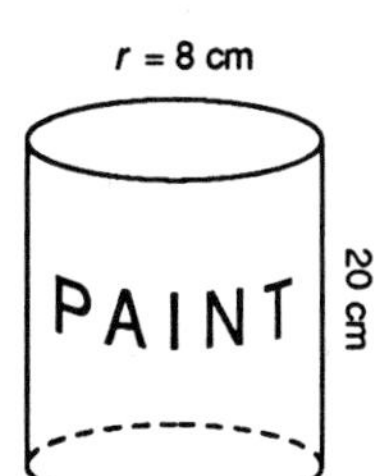

**29.**

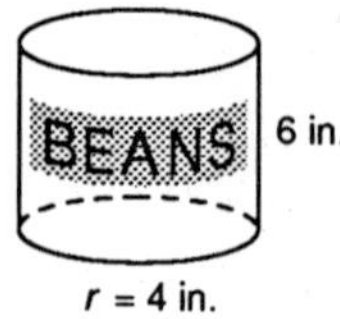


**30.**

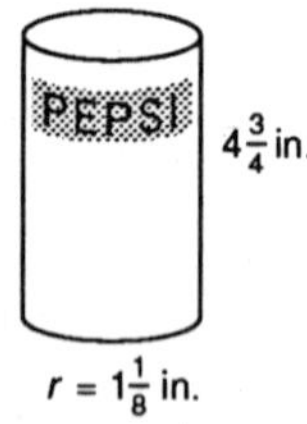


*Find the area of the shaded portions of the figures in Problems 31–36, and then write the expression for the area in factored form.*

**31.**

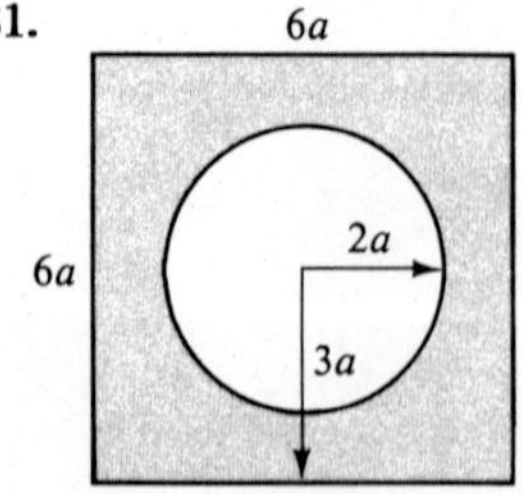


**32.**

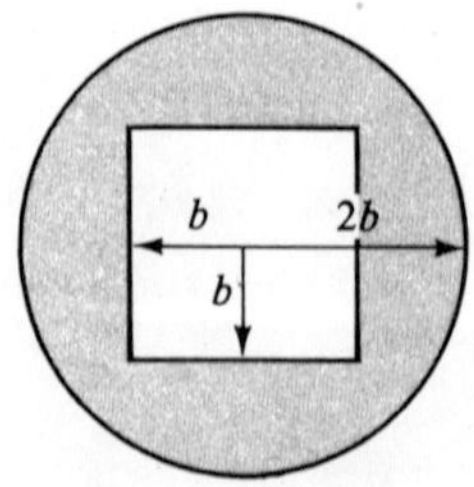


**33.**

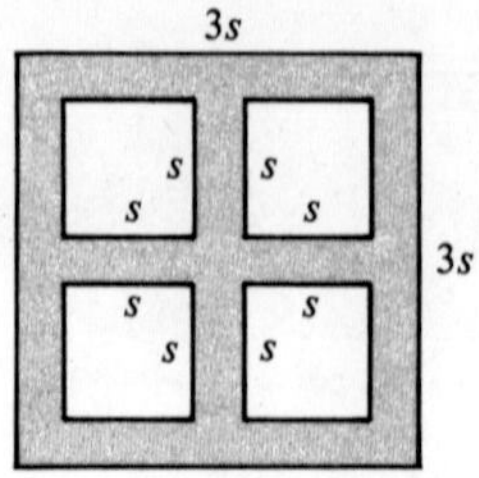


**34.**

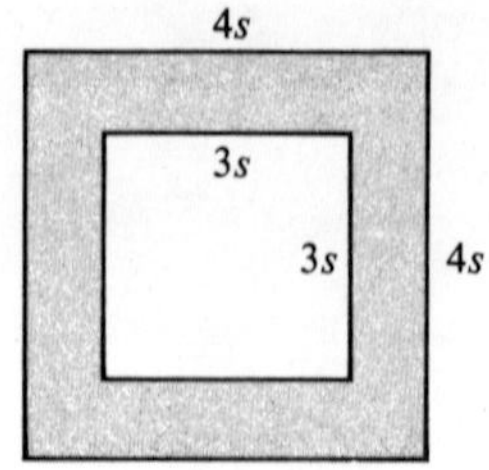


**35.**

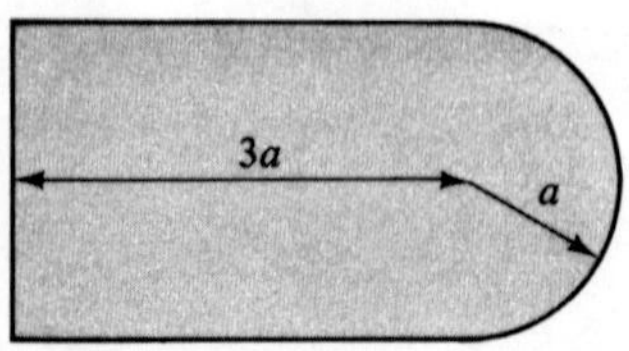


**36.**

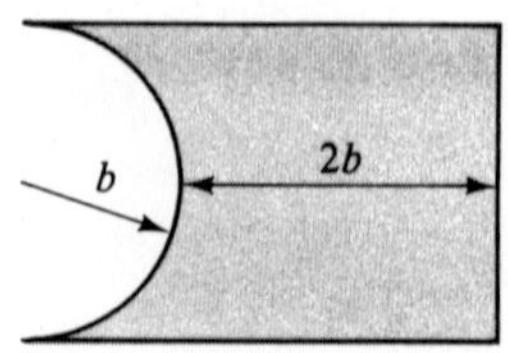


**37.** Which property costs less per square foot?

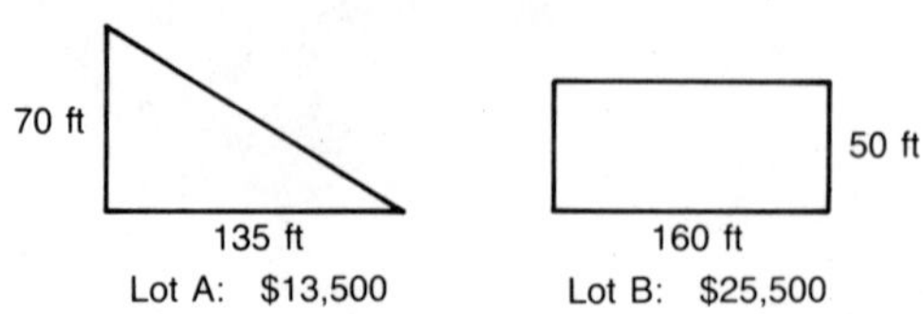


**38.** Which property costs less per square foot?

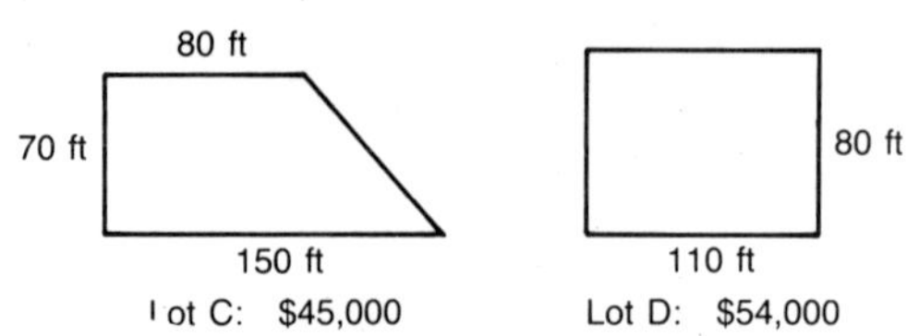

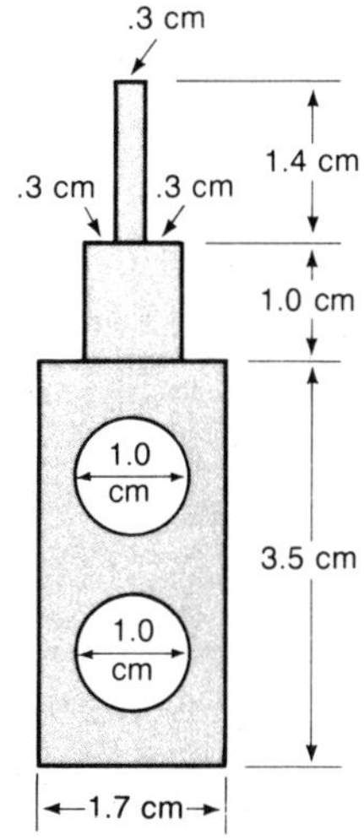


39. If a rectangular piece of property is 750 ft by 1,290 ft, what is the acreage?
40. How many square feet are there in $4\frac{1}{2}$ acres?
41. Find the area of the shaded portion of the figure in the margin, correct to the nearest square centimeter.
42. If the area of a right triangle labeled $\triangle ABC$ (with right angle at $C$) is 12, and if the length of $BC$ is 9, then what is length of $AC$?
43. A circular garden has an area of $800\pi$ ft$^2$. What is its radius? (Give the exact answer.)
44. A square plot of land is partitioned into three rectangular lots. If the area of the entire plot of land is 6,400 ft$^2$, what is the area of the lot shaded in the figure at the right?

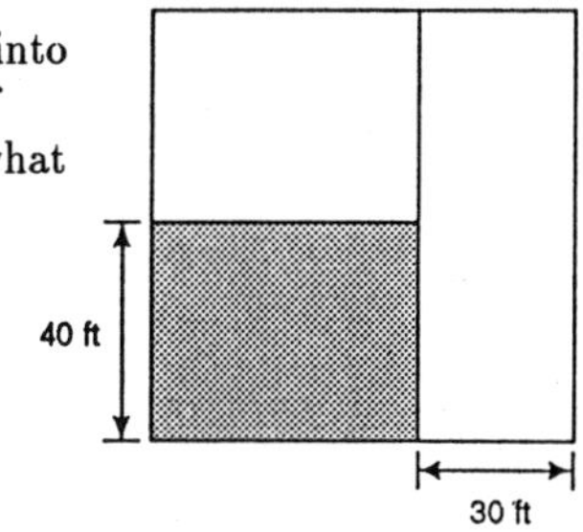


45. If a living room measures 22 ft by 13 ft and needs to be carpeted with a carpet costing \$52 per square yard (including labor and pad), what is the total cost for carpeting this living room? Assume that you cannot purchase part of a square yard.
46. What is the cost of seeding a rectangular lawn 100 ft by 30 ft if 1 pound of seed costs \$5.85 and covers 150 square feet?
47. A certain tile imported from Italy is 25 cm$^2$. In order to determine the number of tiles (which cost \$.85 each), you measure your room using the metric system and find the dimensions to be 5 m by 3.8 m. How much will the tile for this room cost? (Note: 10,000 cm$^2$ = 1 m$^2$.)
48. A small pizza (10-in. diameter) sells for \$7.95, and a large pizza (14 in.) sells for \$13.95. Which pizza is the better price?
49. What is the area of a dollar bill?
50. What is the area of a VISA credit card?
51. Find the area of a circle whose circumference is $8\pi$.
52. What happens to the area of a circle if its radius is doubled?
53. Find the area of a washer if the radius of the larger circle is 10 in. and the area of the smaller circle is 9 in.$^2$.
54. Given a circle of radius 8 in. inscribed in a square, what is the perimeter of the square?

# CHAPTER 4 MEASURING VOLUME

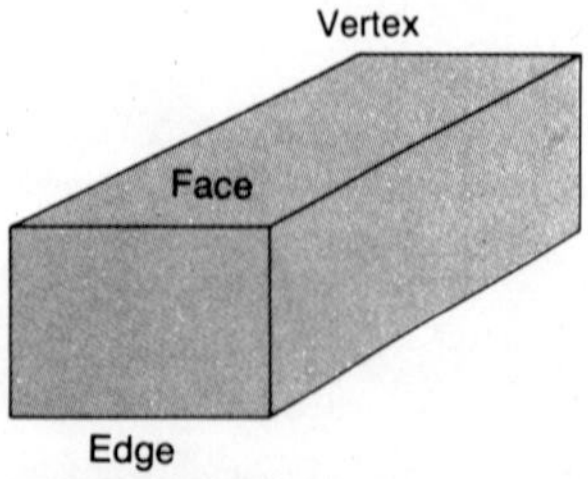


Figure 22 A box, a rectangular solid

### 4.1 Regular Polyhedra

In the last section, we found the surface area of a box. The next reasonable question is, "How much space is inside the box?" A box is made up of *faces*, which meet at *edges*; the "corners" of the box are called *vertices* (plural of *vertex*). The three-dimensional box shown in Figure 22 is one member of a group of objects called **polyhedra.** Polyhedra have faces that are made up entirely of polygons.

Figure 23 shows the five polyhedra whose component polygons are identical in size and shape. Such polygons are called *regular*, and thus the polyhedra in Figure 23 are called **regular polyhedra.** These are the only five regular polyhedra; there are no others.

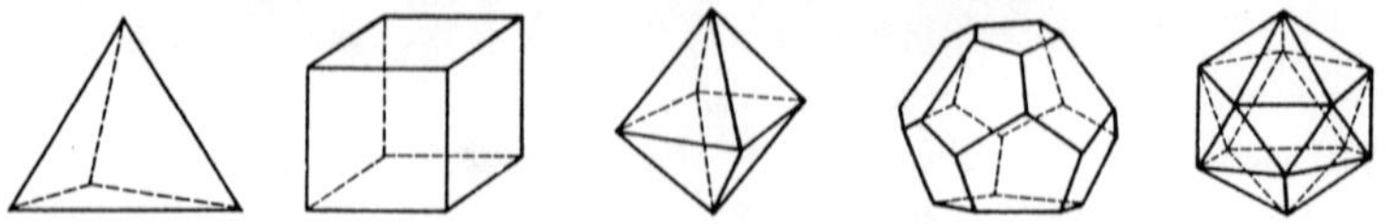

| Tetrahedron (4 faces) | Hexahedron (cube; 6 faces) | Octahedron (8 faces) | Dodecahedron (12 faces) | Icosahedron (20 faces) |
|---|---|---|---|---|

Figure 23 Regular polyhedra

### 4.2 Volume of a Rectangular Solid

We begin our study of volume by looking at the volume of a cube.

**VOLUME OF A CUBE**

The volume $V$ of a cube with edge $s$ is

$$V = s^3$$

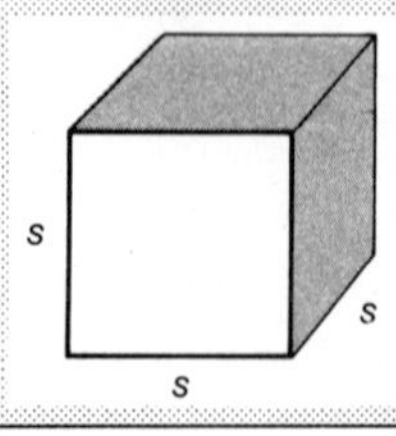


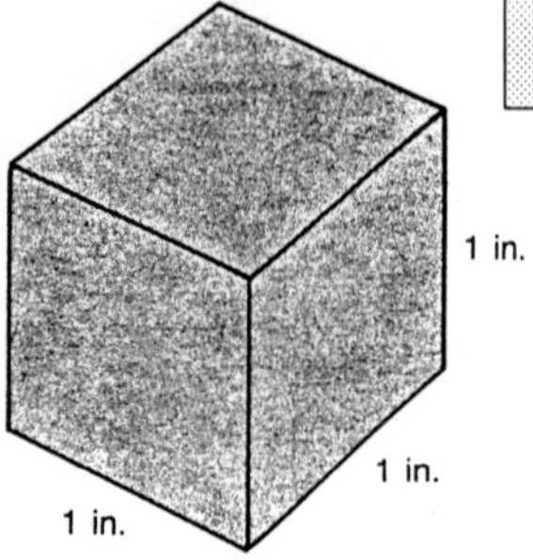


A cubic inch and a cubic centimeter (shown in Figure 24) are used as the basis for volume measurement.

1 cubic inch
Abbreviated 1 cu in. or 1 $in.^3$

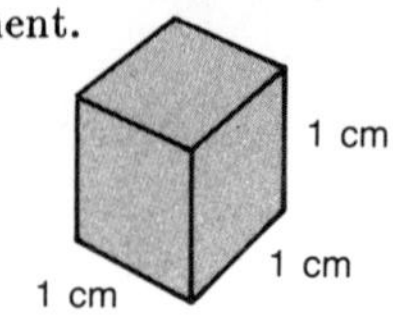


1 cubic centimeter
Abbreviated 1 cu cm or 1 $cm^3$

Figure 24 Basic units of measurement for measuring volume

**EXAMPLE 1**

Find the volume of a box 4 in. by 6 in. by 4 in.

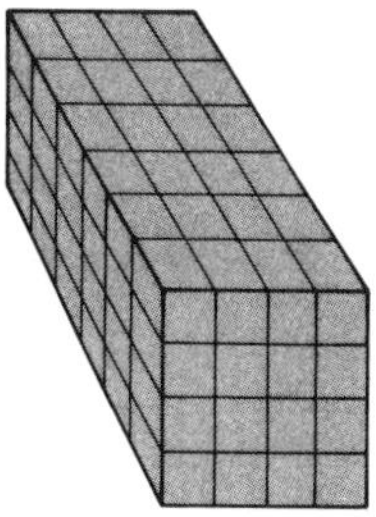

*Solution*

There are 24 cubic inches on the bottom layer of cubes. Do you see how many layers of cubes will fill the solid? Since there are four layers with 24 cubes in each, the total is 96 in.$^3$ You can find this by multiplication:

$4 \times 6 \times 4 = 96$

□

**VOLUME OF A RECTANGULAR SOLID**

The volume $V$ of a rectangular solid (a box) with edges $l$, $w$, and $h$ is

$$V = lwh$$

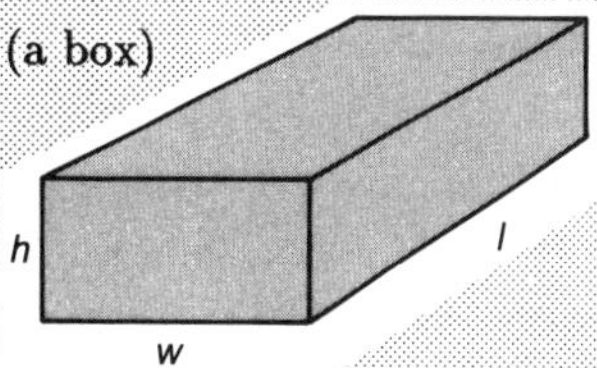


**EXAMPLE 2**

Find the volume of each solid.

*Solution*

**a.**

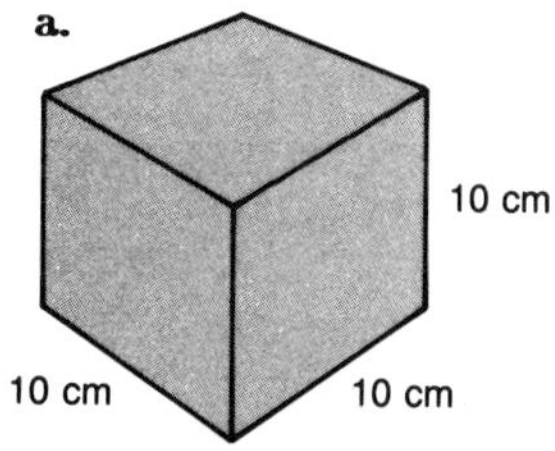


$$\begin{aligned} V &= s^3 \\ &= (10 \text{ cm})^3 \\ &= 1{,}000 \text{ cm}^3 \end{aligned}$$

**b.**

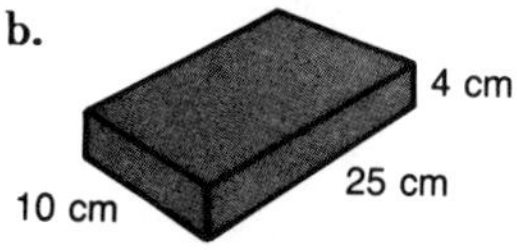


$$\begin{aligned} V &= lwh \\ &= (25 \text{ cm})(10 \text{ cm})(4 \text{ cm}) \\ &= 1{,}000 \text{ cm}^3 \end{aligned}$$

**c.**

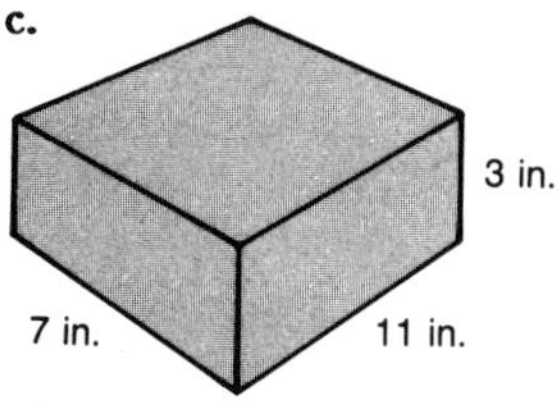


$$\begin{aligned} V &= lwh \\ &= (11 \text{ in.})(7 \text{ in.})(3 \text{ in.}) \\ &= 231 \text{ in.}^3 \end{aligned}$$

□

### 4.3 Volume of Other Solids

From time to time, you may have to find the volume of other solids. If $B$ is the area of the base of the solid, many of the formulas in the following box are similar.

## VOLUME OF VARIOUS SOLIDS

Right prism
$V = Bh$

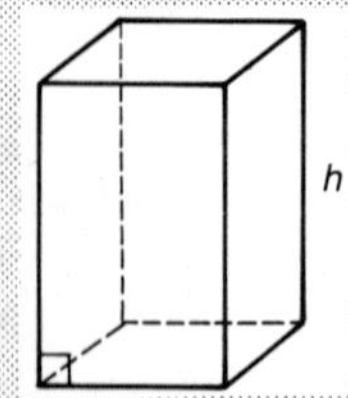


Right circular cylinder
$V = Bh$

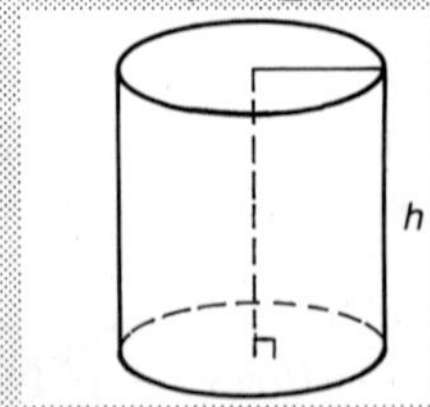


Sphere
$V = \frac{4}{3}\pi r^3$

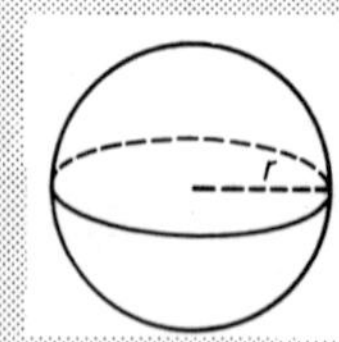


Pyramid
$V = \frac{1}{3}Bh$

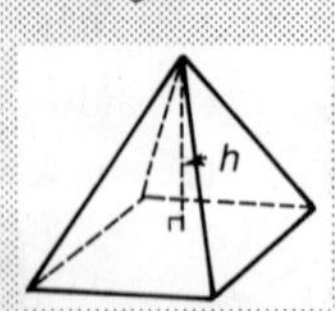


Right circular cone
$V = \frac{1}{3}Bh$

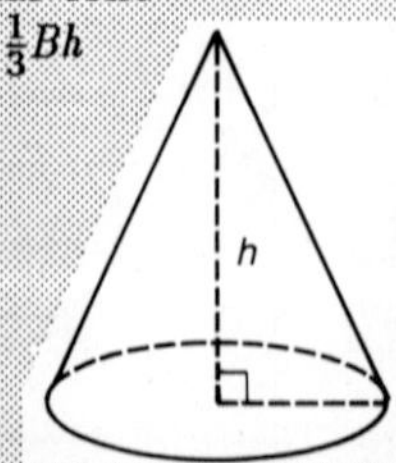


**EXAMPLE 3** Find the volume of the figure at right (rounded to the nearest hundred).

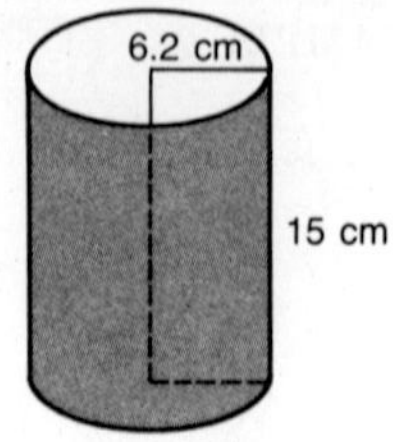


*Solution* This is a right circular cylinder.

$$\begin{aligned} V &= Bh \\ &= \pi(6.2)^2(15) \\ &\approx 1{,}811.4423 \end{aligned}$$

The volume is 1,800 $\text{cm}^3$. □

**EXAMPLE 4** Find the volume of the figure at right (rounded to the nearest cubic foot).

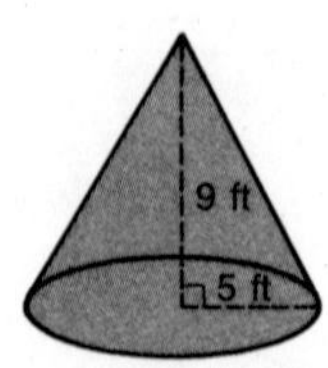


*Solution* This is a right circular cone.

$$\begin{aligned} V &= \tfrac{1}{3}Bh \\ &= \tfrac{1}{3}\pi r^2 h \\ &= \tfrac{1}{3}\pi(5)^2 9 \\ &= 75\pi \\ &\approx 235.61945 \end{aligned}$$

*Notice that* $B = \pi r^2$

The volume is 236 $\text{ft}^3$. □

**EXAMPLE 5** Find the volume of the figure at right (rounded to the nearest 10 cubic feet).

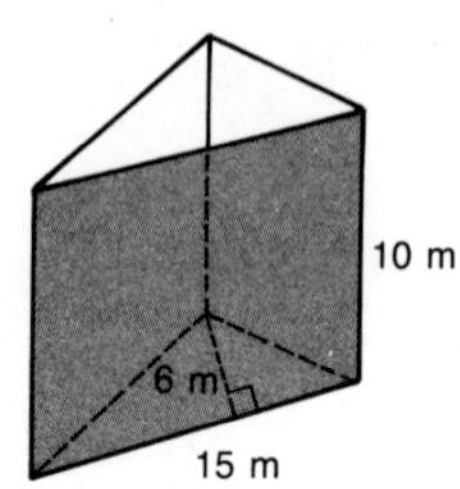


*Solution* This is a right triangular prism.

$$\begin{aligned} V &= Bh \\ &= \tfrac{1}{2}bah \\ &= \tfrac{1}{2}(15)(6)(10) \end{aligned}$$

*Notice that* $B = \frac{1}{2}ba$

$= 450$

The volume is 450 in.$^3$ □

**EXAMPLE 6** Find the volume of the figure at right (rounded to the nearest hundredth).

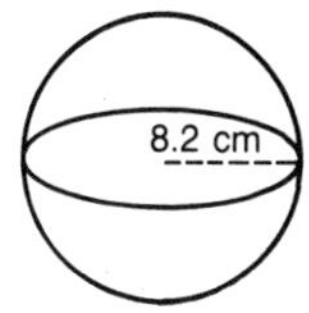


*Solution* This is a sphere.

$$V = \tfrac{4}{3}\pi r^3$$
$$= \tfrac{4}{3}\pi(8.2)^3$$
$$\approx 2{,}309.5649$$

The volume is 2,300 cm$^3$. □

**EXAMPLE 7** Find the volume of the figure at right (rounded to the nearest unit).

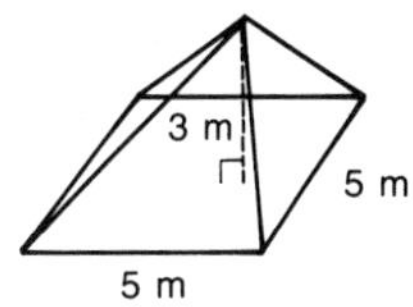


*Solution* This is a pyramid.

$$V = \tfrac{1}{3}Bh$$
$$= \tfrac{1}{3}(5)^2(3)$$
$$= \tfrac{25}{3}(3)$$
$$= 25$$

The volume is 25 m$^2$. □

### 4.4 Converting Units

Sometimes the dimensions for the volume you are finding are not given in the same units. In such cases, you must convert all units to be the same. The common conversions are as follows:

| | |
|---|---|
| 1 ft = 12 in. | To convert feet to inches, multiply by 12; inches to feet, divide by 12. |
| $(1\text{ ft})^2 = (12\text{ in.})^2$ | To convert square feet to square inches, multiply by 144; |
| $1\text{ ft}^2 = 144\text{ in.}^2$ | square inches to square feet, divide by 144. |
| $(1\text{ ft})^3 = (12\text{ in.})^3$ | To convert cubic feet to cubic inches, multiply by 1,728; |
| $1\text{ ft}^3 = 1{,}728\text{ in.}^3$ | cubic inches to cubic feet, divide by 1,728. |
| 1 yd = 3 ft | To convert yards to feet, multiply by 3; feet to yards, divide by 3; |
| $(1\text{ yd})^2 = (3\text{ ft})^2$ | To convert square yards to square feet, multiply by 9; |
| $1\text{ yd}^2 = 9\text{ ft}^2$ | square feet to square yards, divide by 9. |
| $(1\text{ yd})^3 = (3\text{ ft})^3$ | To convert cubic yards to cubic feet, multiply by 27; |
| $1\text{ yd}^3 = 27\text{ ft}^3$ | cubic feet to cubic yards, divide by 27. |

**EXAMPLE 8**

Suppose you are pouring a rectangular driveway with dimensions of 24 ft by 65 ft. The depth of the driveway is 3 in., and concrete is ordered by the yard. By a "yard" of concrete, we mean a cubic yard. You cannot order part of a yard of concrete. How much concrete should you order?

*Solution*

There are three different measurements in this problem: inches, feet, and yards. Since we want the answer in cubic yards, we will convert all of these measurements to yards:

$24 \text{ ft} = (24 \div 3) \text{ yd} = 8 \text{ yd}$

$65 \text{ ft} = (65 \div 3) \text{ yd} = \frac{65}{3} \text{ yd}$

$3 \text{ in.} = (3 \div 36) \text{ yd} = \frac{3}{36} \text{ yd} = \frac{1}{12} \text{ yd}$

The calculation is now carried out in cubic yards:

$$\begin{aligned} V &= lwh \\ &= \tfrac{65}{3}(8)(\tfrac{1}{12}) \text{ yd}^3 \\ &\approx 14.444444 \text{ yd}^3 \quad \textit{By calculator} \end{aligned}$$

You must order 15 yards of concrete. □

## 4.5 PROBLEM SET 4

**A**

*In Problems 1–3, find the volume of each solid by counting the number of cubic centimeters in each box.*

**1.**

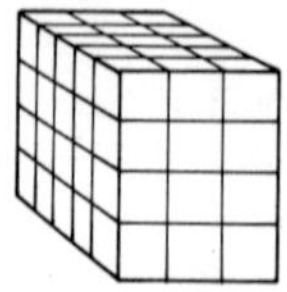

**2.**

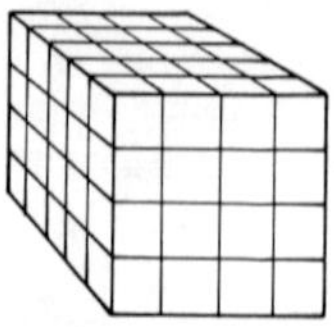

**3.**

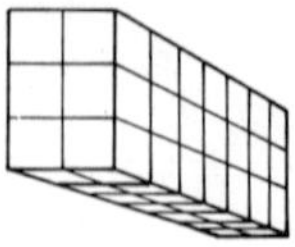

*Find the volume of each solid in Problems 4–13. Assume that the measurements are exact, and do not round your answers.*

**4.**

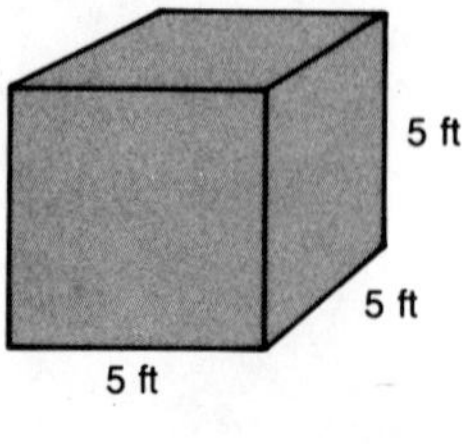


**5.**

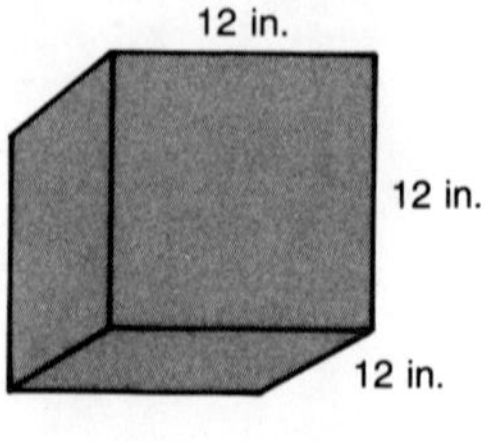


**6.**

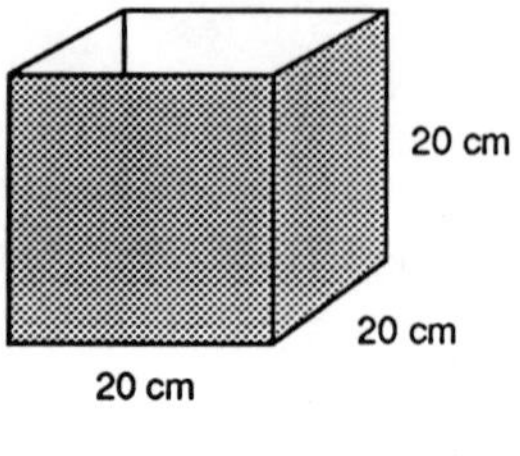


**7.**

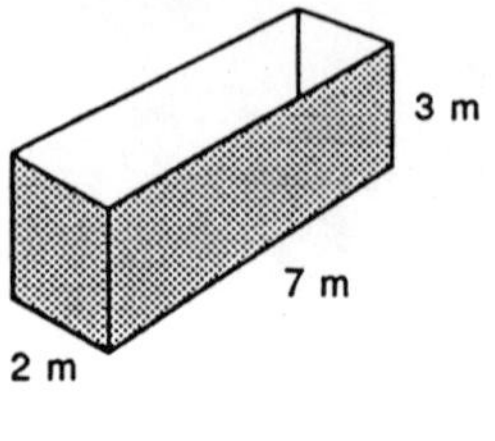

**8.**

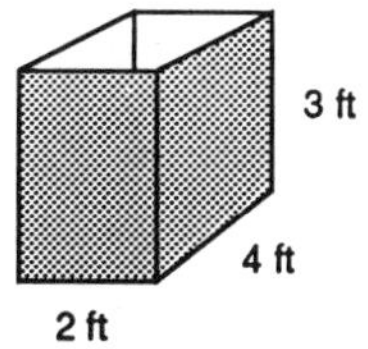


**9.**

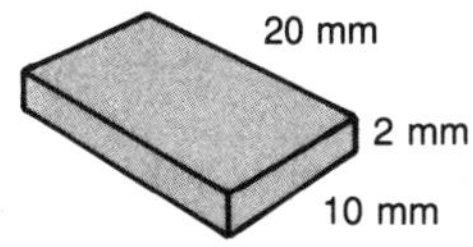


**10.**

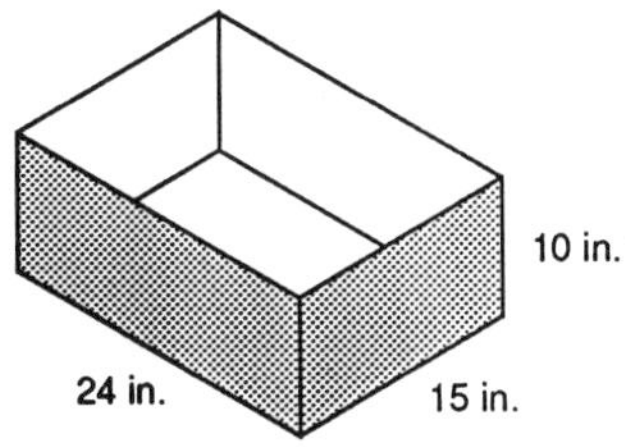


**11.**

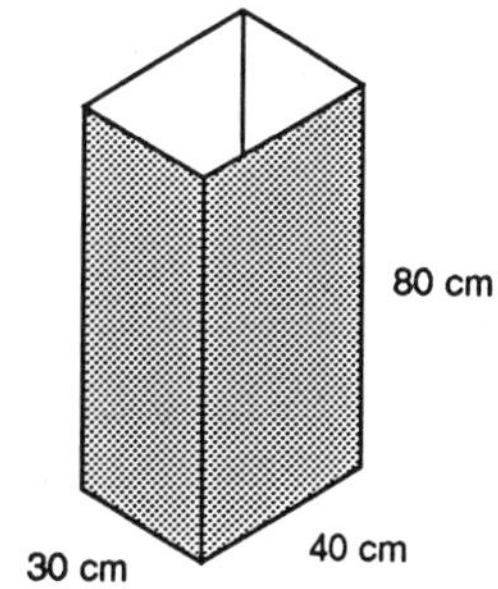


**12.**

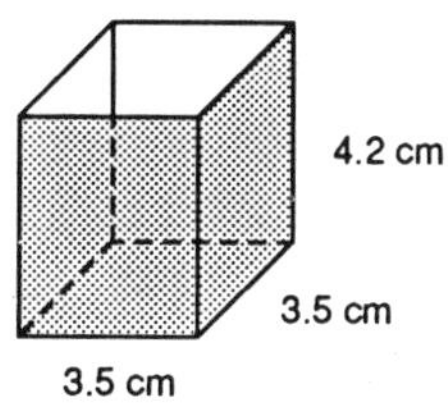


**13.**

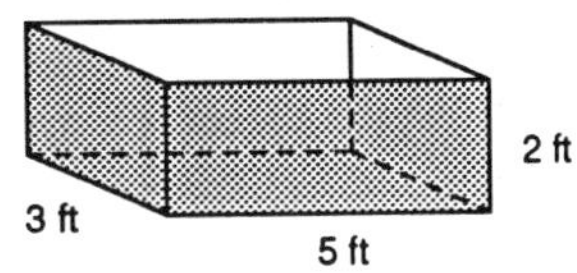


## B

*Find the volumes of the solids in Problems 14–19 to the nearest cubic unit.*

**14.**

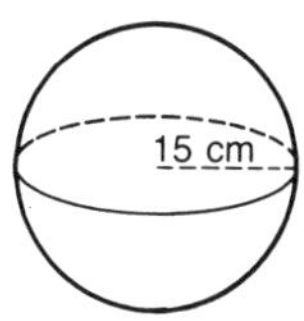


**15.**

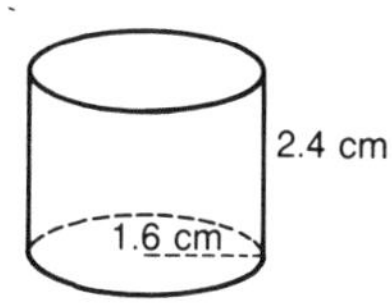


**16.**

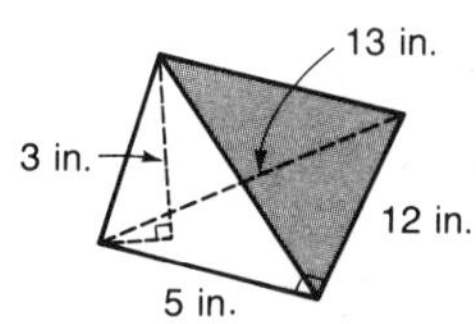


**17.**

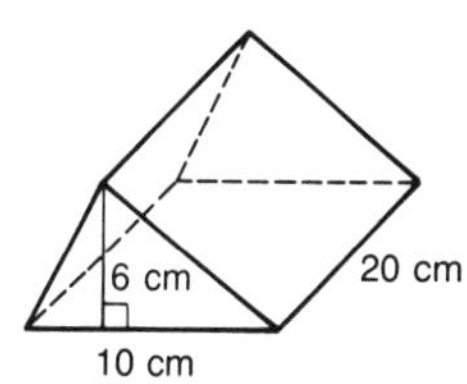


**18.**

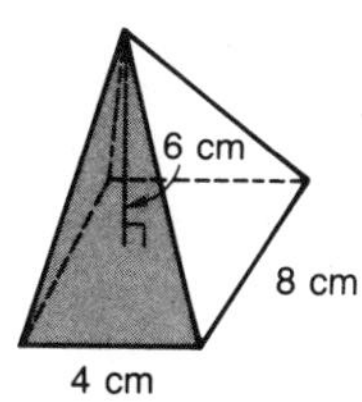


**19.**

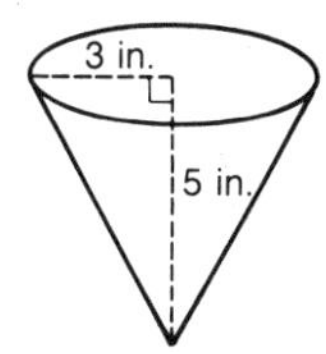


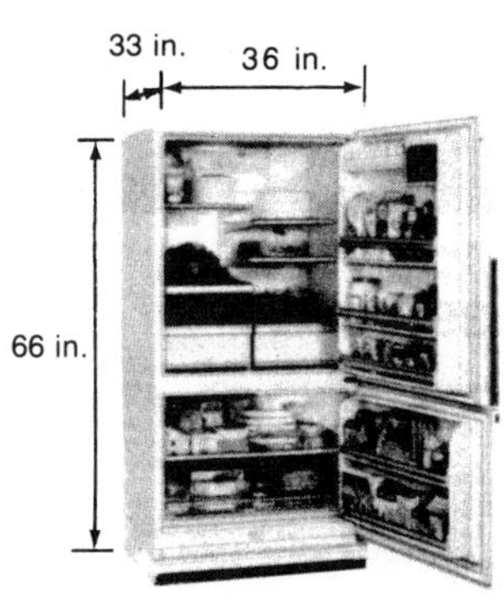


**20.** The exterior dimensions of a refrigerator are as shown in the margin.

**a.** How many cubic feet are contained within the refrigerator?

**b.** If it is advertised as a 19-cu-ft refrigerator, how much space is taken up by the motor, insulation, and so on?

**21.** The exterior dimensions of a freezer are 48 inches by 36 inches by 24 inches, and it is advertised as being 27.0 ft$^3$. Is the advertised volume correctly stated?

22. In the figure shown at right, two rectangular solids meet to form the *L*-shaped solid. What is the volume?

23. Suppose you must order concrete for a sidewalk 50 ft by 4 ft by 4 in. deep. How much concrete is required? (Answer to the nearest half cubic yard.)

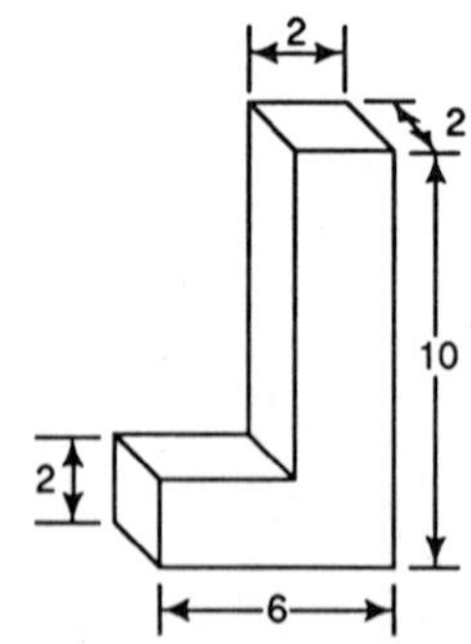


*Use the plot plan shown in the figure below to answer Problems 24–28. Answer to the nearest half cubic yard.**

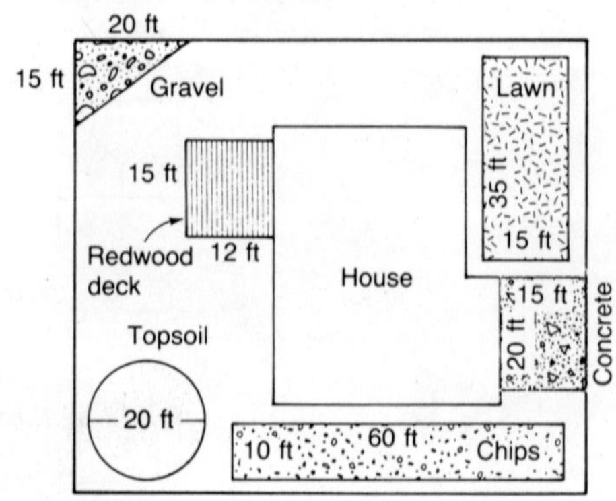


24. How many cubic yards of sawdust are needed for preparation of the lawn area if it is to be spread to a depth of 6 inches?
25. How many cubic yards of chips are necessary if they are to be placed to a depth of 3 inches?
26. How much topsoil needs to be hauled in if it is to be spread to a depth of 12 inches?
27. How much gravel is necessary if it is to be laid to a depth of 3 inches?
28. Suppose you wish to pave the area labeled "concrete." How much concrete is needed if it is to be poured to a depth of 4 inches?

*In practice, you would not round to the nearest half yard; instead, you would round up, to ensure that you got enough material. However, for consistency in this book, we will round using the usual mathematical rules.

# CHAPTER 5 CONGRUENT AND SIMILAR TRIANGLES

Congruence and similarity are two of the most important concepts in geometry. These concepts are used with lines, angles, triangles, and other geometric figures that have something in common.

### 5.1 Congruent Triangles

We say that two triangles $ABC$ and $DEF$ are **congruent** if they have the same size and shape. We denote this by writing $\triangle ABC \simeq \triangle DEF$. From this symbolic statement we conclude that all six corresponding parts are congruent. We call these **corresponding parts.**

EXAMPLE 1

Name the corresponding parts of the given triangles.

**a.** $\triangle ABC \simeq \triangle A'B'C'$ **b.** $\triangle RST \simeq \triangle UST$

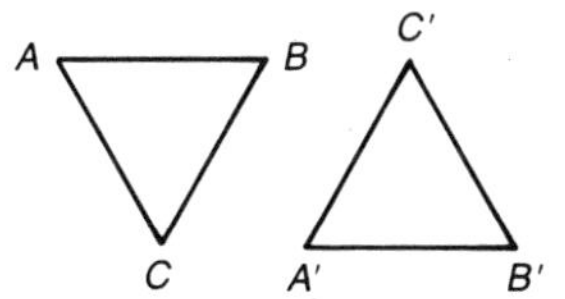


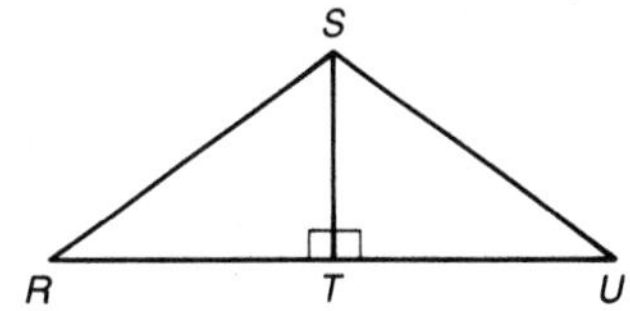


*Solution*

**a.**
$AB$ corresponds to $A'B'$
$AC$ corresponds to $A'C'$
$BC$ corresponds to $B'C'$
$\angle A$ corresponds to $\angle A'$
$\angle B$ corresponds to $\angle B'$
$\angle C$ corresponds to $\angle C'$

**b.**
$RS$ corresponds to $US$
$RT$ corresponds to $UT$
$ST$ corresponds to $ST$
$\angle R$ corresponds to $\angle U$
$\angle RTS$ corresponds to $\angle UTS$
$\angle RST$ corresponds to $\angle UST$

□

In Chapter 1, we defined the word **congruent** to mean angles having the same size and shape. Line segments, angles, triangles, or other geometric figures are *congruent* if they have the same size and shape. In this section, we focus on triangles.

**CONGRUENT TRIANGLES**

> Two triangles are **congruent** if their corresponding sides have the same length and their corresponding angles have the same measure.

In order to prove that two triangles are congruent, you must show that they have the same size and shape. It is not necessary to show that all six parts (three sides and three angles) are congruent; if some of these six parts are congruent, it necessarily follows that the other parts are congruent. Three important properties are used to show triangle congruence:

**CONGRUENT TRIANGLE PROPERTIES**

> **SIDE-SIDE-SIDE (SSS)**
> If three sides of one triangle are congruent to three sides of another triangle, then the two triangles are congruent.
>
> **SIDE-ANGLE-SIDE (SAS)**
> If two sides of one triangle and the angle between those sides are congruent to the corresponding sides and angle of another triangle, then the two triangles are congruent.
>
> **ANGLE-SIDE-ANGLE (ASA)**
> If two angles and the side that connects them on one triangle are congruent to the corresponding angles and side of another triangle, then the two triangles are congruent.

**EXAMPLE 2** Determine if each pair of triangles is congruent. If so, cite one of the congruent triangle properties.

*Solution*

**a.**

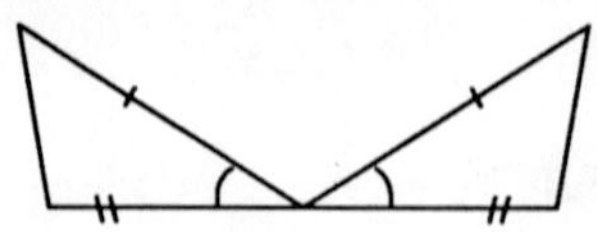

Congruent; SAS

**b.**

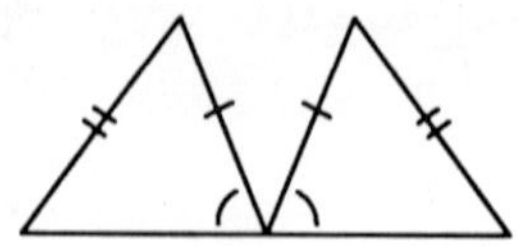

Not congruent

**c.**

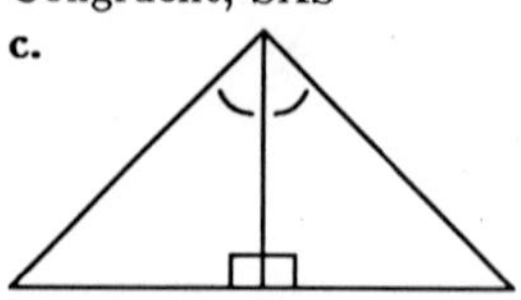

Congruent; ASA

**d.**

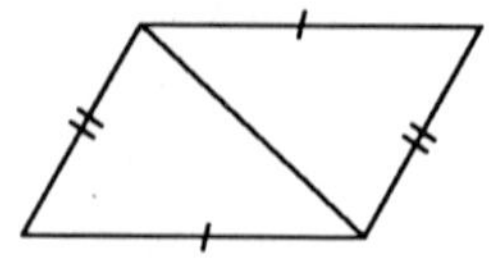

Congruent; SSS

A side that is in common to two triangles obviously is equal in length to itself, and it does not need to be marked. □

In geometry, the main use of congruent triangles is when we want to know whether an angle from one triangle is congruent to an angle from a different triangle, or when we want to know whether a side from one triangle is the same length as the side from another triangle. In order to do this, we often prove that one triangle is congruent to the other (using one of the three congruent triangle properties) and then use the following property:

*Corresponding parts of congruent triangles are congruent.*

### 5.2 Similar Triangles

It is possible for two figures to have the same shape, but not necessarily the same size. (Notice that congruent triangles must also be similar, but similar triangles are not necessarily congruent.) These figures are called **similar** figures. We will now focus on similar **triangles.** If $\triangle ABC$ is similar to $\triangle DEF$, we write

$\triangle ABC \sim \triangle DEF$

Similar triangles are shown in Figure 25.

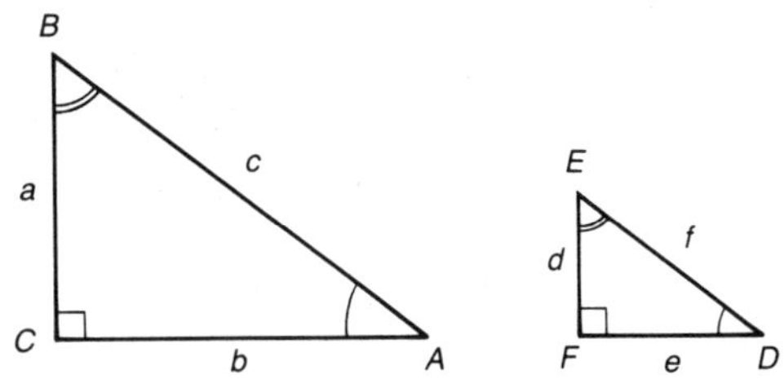


Figure 25 Similar triangles

You should note that congruent triangles must be similar, but similar triangles are not necessarily congruent. Since similar figures have the same shape, we talk about **corresponding angles** and **corresponding sides.** The corresponding angles of similar triangles are the angles that have the same measure. It is customary to label the vertices of triangles with capital letters and the sides opposite the angles at those vertices with corresponding lower-case letters. It is easy to see that, if the triangles are similar, the corresponding sides are the sides opposite equal angles. In Figure 25 we see that

$m\angle A = m\angle D$ so these are corresponding angles;
$m\angle B = m\angle E$ so these are corresponding angles; and
$m\angle C = m\angle F$ so these are corresponding angles.
Side $a$ ($BC$) is opposite $\angle A$, and $d$ ($EF$) is opposite $\angle D$, so we say that $a$ corresponds to $d$;
$b$ corresponds to $e$; and
$c$ corresponds to $f$.

Even though corresponding angles are equal, corresponding sides do not need to have the same length. If they do have the same length, then the triangles are congruent. However, when they are not the same length, we can say they are proportional. As Figure 25 illustrates, when we say the sides are proportional, we mean

$$\frac{a}{b} = \frac{d}{e} \qquad \frac{a}{c} = \frac{d}{f} \qquad \frac{b}{c} = \frac{e}{f}$$

$$\frac{b}{a} = \frac{e}{d} \qquad \frac{c}{a} = \frac{f}{d} \qquad \frac{c}{b} = \frac{f}{e}$$

**SIMILAR TRIANGLES**

Two triangles are **similar** if two angles of one triangle have the same measure as two angles of the other triangle. If the triangles are similar, then their corresponding sides are proportional.

This property holds for any two figures that are similar, not only for triangles.

**EXAMPLE 3**

Identify pairs of triangles that are similar in Figure 26.

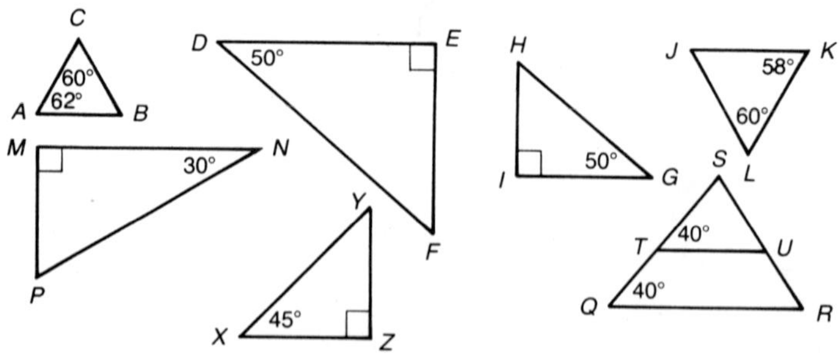


Figure 26

*Solution*

$\triangle ABC \sim \triangle JKL$; $\triangle DEF \sim \triangle GIH$; $\triangle SQR \sim \triangle STU$. □

**EXAMPLE 4**

Given the similar triangles below, find the unknown lengths marked $b'$ and $c'$.

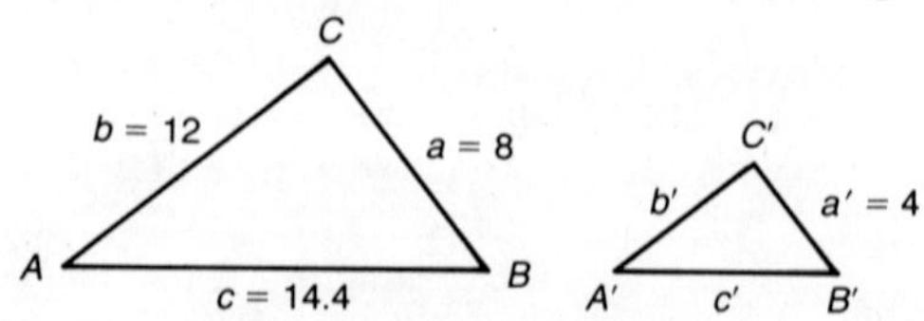


*Solution*

Since corresponding sides are proportional (other proportions are possible), we have

$$\frac{a'}{a} = \frac{b'}{b} \qquad \frac{a}{c} = \frac{a'}{c'}$$

$$\frac{4}{8} = \frac{b'}{12} \qquad \frac{8}{14.4} = \frac{4}{c'}$$

$$b' = \frac{4(12)}{8} \qquad c' = \frac{14.4(4)}{8}$$

$$= 6 \qquad = 7.2$$

□

**EXAMPLE 5**

In equilateral $\triangle ABC$, suppose $DE = 2$ and is parallel to $AB$ as shown at right. If $AB$ is three times as long as $DE$, what is the perimeter of quadrilateral $ABED$?

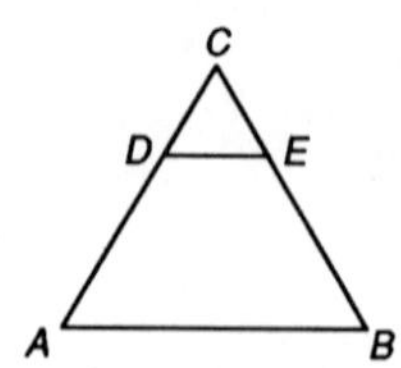


*Solution*

$\triangle ABC \sim \triangle DEC$ so $\triangle DEC$ is equilateral. This means that $CE$ and $DC$ both have length 2; thus, $EB$ and $AD$ both have length 4. The perimeter of the quadrilateral is

$$AB + BE + DE + AD = 6 + 4 + 2 + 4 = 16$$

□

Finding similar triangles is simplified even further if we know the triangles are right triangles, because then the triangles are similar if one of the acute angles has the same measure as an acute angle of the other.

**EXAMPLE 6**

Suppose that a tree and a yardstick are casting shadows as shown in Figure 27. If the shadow of the yardstick is 3 yards long and the shadow of the tree is 12 yards long, use similar triangles to estimate the height of the tree if you know that $m\angle S = m\angle S'$.

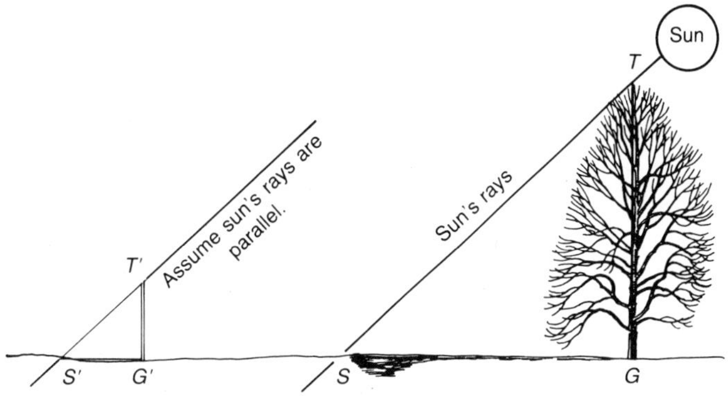


**Figure 27**

*Solution*

Since $\angle G$ and $\angle G'$ are right angles, and since $m\angle S = m\angle S'$, we see that $\triangle SGT \sim \triangle S'G'T'$. Therefore, corresponding sides are proportional.

$$\frac{1}{3} = \frac{h}{12}$$
$$h = \frac{1(12)}{3}$$
$$= 4$$

The tree is 4 yards tall. □

### 5.3 Right Triangles

There is a relationship between the size of the angles of a right triangle and the ratios of the lengths of the sides. In a right triangle, the side opposite the right angle is called the **hypotenuse.** Each of the acute angles of a right triangle has one side that is the hypotenuse. The other side of that angle is called the **adjacent side.**

EXAMPLE 7

In $\triangle ABC$ with right angle at $C$,

the hypotenuse is $c$,

the side adjacent to $\angle A$ is $b$,
the side opposite $\angle A$ is $a$,

the side adjacent to $\angle B$ is $a$, and
the side opposite $\angle$B is $b$. □

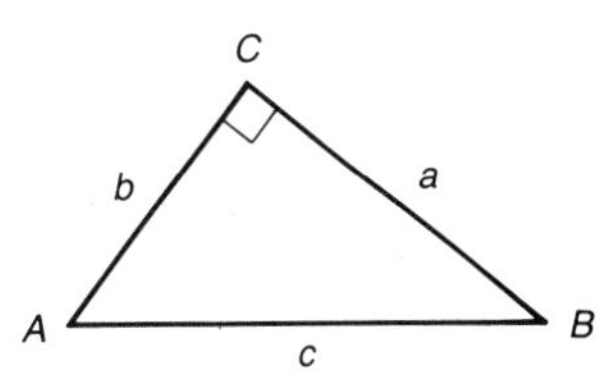


**Pythagorean Theorem**

An important theorem about right triangles is the Pythagorean Theorem.

PYTHAGOREAN THEOREM

> For any right triangle with sides $a$ and $b$ and hypotenuse $c$,
>
> $$a^2 + b^2 = c^2$$

If $c$ is not known, you can use the square root property from algebra to find $c$ (disregard the negative value because $c$ is a distance):

$$c = \sqrt{a^2 + b^2}$$

If either $a$ or $b$ is not known, first solve for that value before using the square root property:

$$a^2 = c^2 - b^2 \qquad b^2 = c^2 - a^2$$

$$a = \sqrt{c^2 - b^2} \qquad b = \sqrt{c^2 - a^2}$$

**EXAMPLE 8** A carpenter wants to make sure that the corner of a room is square (a right angle). If she measures out sides of 3 feet and 4 feet, how long should she make the diagonal (hypotenuse)?

*Solution* The hypotenuse is the unknown, so use the Pythagorean Theorem:

$$\begin{aligned} a^2 + b^2 &= c^2 \\ \sqrt{a^2 + b^2} &= c \\ c &= \sqrt{3^2 + 4^2} \\ &= \sqrt{25} \\ &= 5 \end{aligned}$$

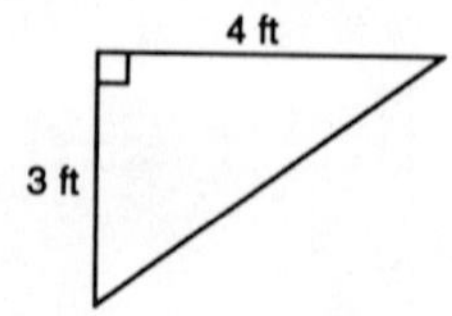


She should make the diagonal 5 feet long. □

**EXAMPLE 9** If a 10-ft ladder is placed against a building so that the base of the ladder is 4 ft away from the building, how high up does the ladder reach?

*Solution* Consider the figure at right. Since $h$ is one of the sides of a right triangle, we can use the Pythagorean Theorem:

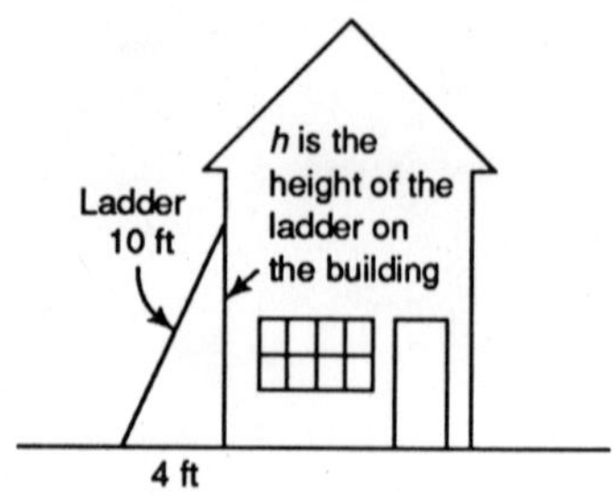


$$\begin{aligned} h &= \sqrt{10^2 - 4^2} \\ &= \sqrt{100 - 16} \\ &= \sqrt{84} \\ &\approx 9.1651514 \end{aligned}$$

The ladder will reach about 9 ft up the side of the building. □

**EXAMPLE 10** What is the ratio of the length of a square's diagonal to its perimeter?

*Solution* First, find the diagonal of a square with side $s$ by using the Pythagorean Theorem:

$$(\text{DIAGONAL})^2 = s^2 + s^2$$

$$\text{DIAGONAL} = \sqrt{2s^2} = s\sqrt{2}$$ *Square root property (disregard the negative values since the diagonal is a length)*

Since a ratio is a fraction, we have:

$$\frac{\text{LENGTH OF DIAGONAL}}{\text{PERIMETER}} = \frac{s\sqrt{2}}{4s} = \frac{\sqrt{2}}{4}$$ □

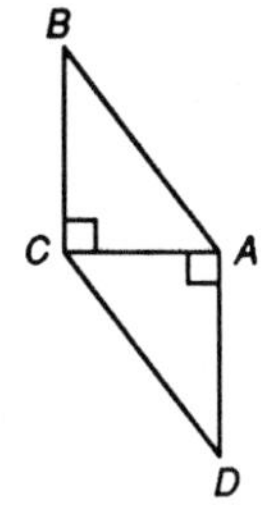


**EXAMPLE 11** In the figure shown at right, $\triangle ABC \simeq \triangle CDA$. If $AB = 6$ and $AC = 4$, what is $AD$?

*Solution* From the Pythagorean Theorem in $\triangle ABC$,

$$BC = \sqrt{AB^2 - AC^2} = \sqrt{6^2 - 4^2} = \sqrt{20} = 2\sqrt{5}$$

Since the triangles are congruent and $AD$ corresponds to $BC$, we see

$$AD = 2\sqrt{5}$$ □

**45°-45°-90° triangle**

Two types of special triangles are often used in applications. The first of these is the right triangle with acute angles measuring 45° each. This triangle, also called an **isosceles right triangle**, is shown in Figure 28.

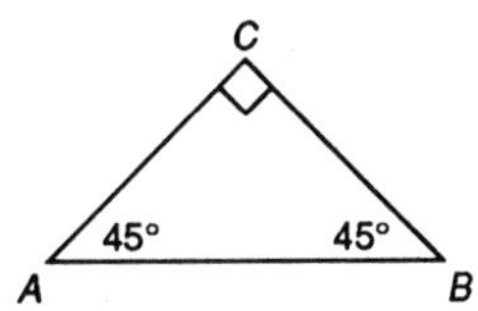


Figure 28

If each leg has length $x$ units, then, by the Pythagorean Theorem,

$$x^2 + x^2 = c^2$$

where $c$ is the hypotenuse. Solve for $c$ to find

$$c = x\sqrt{2}$$

This result is known as the 45°-45°-90° triangle property.

**45°-45°-90° TRIANGLE**

For any right triangle with acute angles measuring 45°:
The legs are congruent, and the hypotenuse has a length equal to $\sqrt{2}$ times the length of one of those legs.

**EXAMPLE 12** Find the length of the hypotenuse in an isosceles right triangle whose leg has length 9.

*Solution* By the 45°-45°-90° property, the hypotenuse has length $9\sqrt{2}$. □

**EXAMPLE 13** Find the length of a leg of an isosceles right triangle if the length of the hypotenuse is 9.

*Solution* By the 45°-45°-90° property, if $x$ is the length of one leg, then

$$x\sqrt{2} = 9$$

$$x = \frac{9}{\sqrt{2}} = \frac{9}{\sqrt{2}} \cdot \frac{\sqrt{2}}{\sqrt{2}} = \frac{9\sqrt{2}}{2} \quad \text{or} \quad \tfrac{9}{2}\sqrt{2}$$ □

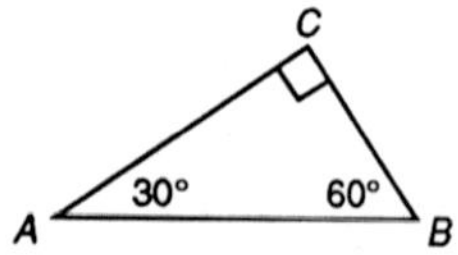


Figure 29

**30°-60°-90° triangle**

A second triangle property involves a right triangle with acute angles measuring 30° and 60°, as shown in Figure 29. In order to derive this property we need to consider two 30°-60°-90° triangles that are joined along their longer legs so that an equilateral triangle can be formed, as shown in Figure 30.

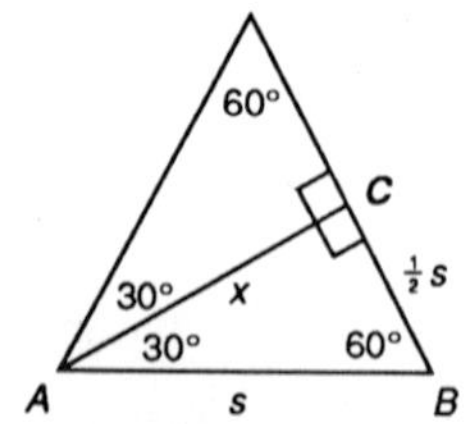


Figure 30

If this equilateral triangle's sides have length $s$, then the shorter leg of the original 30°-60°-90° triangle must have length $\frac{1}{2}s$ because the newly formed triangle is equilateral. This means that if we look at the original 30°-60°-90° triangle, the length of the hypotenuse is twice the length of the shorter leg. To find the length of the longer leg $x$, we will use the Pythagorean Theorem:

$$s = (\tfrac{1}{2}s)^2 + x^2$$

$$s^2 = \tfrac{1}{4}s^2 + x^2$$

$$\tfrac{3}{4}s^2 = x^2$$

$$x = \tfrac{\sqrt{3}}{2}s \qquad \textit{Positive, because } x \textit{ and } s \textit{ are distances}$$

This leads to the following property.

**30°-60°-90° TRIANGLE**

For any right triangle with acute angles measuring 30° and 60°:

1. The hypotenuse is twice as long as the leg opposite the 30° angle (the shorter leg).
2. The leg opposite the 30° angle (the shorter leg) is $\frac{1}{2}$ as long as the hypotenuse.
3. The leg opposite the 60° angle (the longer leg) equals the length of the other (shorter) leg times $\sqrt{3}$.
4. The leg opposite the 30° angle equals the length of the other leg divided by $\sqrt{3}$.

**EXAMPLE 14**

An equilateral triangle has a perimeter of 30. What is the height of this triangle?

*Solution*

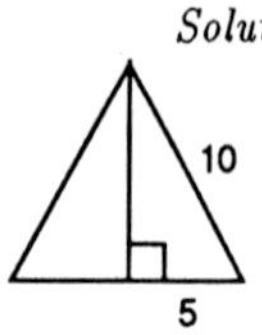


Figure 31

Since the triangle is equilateral, the sides are the same length, so the length of one side is 10. This is also the hypotenuse of a 30°-60°-90° triangle (see Figure 31). The shorter leg of this triangle is 5 (one-half the hypotenuse), and the height of the triangle is the longer leg of the 30°-60°-90° triangle. We know that this is $5\sqrt{3}$. □

### 5.4 Trigonometric Ratios

In a right triangle certain ratios of the sides are important enough to be given specific names. They are called sin $A$ (pronounced "sine $A$", cos $A$ (pronounced "cosine $A$"), and tan $A$ (pronounced "tangent $A$"). These ratios are defined in the following box.

**TRIGONOMETRIC RATIOS**

In a right triangle $ABC$ with right angle at $C$,

$$\sin A = \frac{\text{OPPOSITE SIDE OF } A}{\text{HYPOTENUSE}}$$

$$\cos A = \frac{\text{ADJACENT SIDE OF } A}{\text{HYPOTENUSE}}$$

$$\tan A = \frac{\text{OPPOSITE SIDE OF } A}{\text{ADJACENT SIDE OF } A}$$

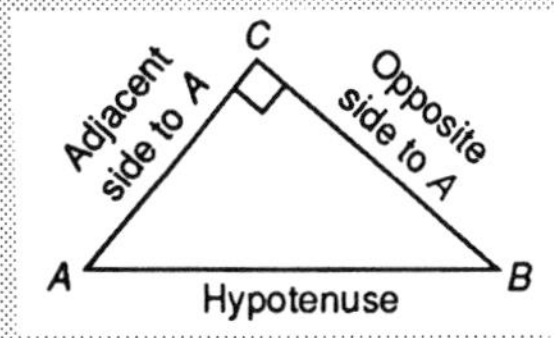


**EXAMPLE 15** Given a right triangle with sides 5 and 12, find the trigonometric ratios for the angles $A$ and $B$. Show your answers in both common fraction and decimal fraction form, with decimals rounded to four places.

*Solution* First use the Pythagorean Theorem to find the length of the hypotenuse:

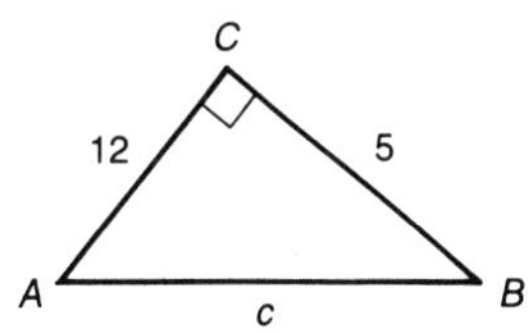


$$\begin{aligned} c &= \sqrt{5^2 + 12^2} \\ &= \sqrt{25 + 144} \\ &= \sqrt{169} \\ &= 13 \end{aligned}$$

$\sin A = \frac{5}{13} \approx .3846$; $\cos A = \frac{12}{13} \approx .9231$; $\tan A = \frac{5}{12} \approx .4167$.

$\sin B = \frac{12}{13} \approx .9231$; $\cos B = \frac{5}{13} \approx .3846$; $\tan B = \frac{12}{5} = 2.4$. □

Tables of ratios for different angles are available (see Table II in Appendix A). Certain calculators have keys for the sine, cosine, and tangent ratios. If you have such a calculator, you should press the angle first and then the key for the trigonometric ratio.

**EXAMPLE 16** Use either Table II or a calculator to find the trigonometric ratios for

**a.** sin 45° **b.** cos 32° **c.** tan 19°

*Solution*

**a.** sin 45° = .7071 by table; by calculator, press [45] [sin]: 0.7071067

**b.** cos 32° = .8480 by table; by calculator, press [32] [cos]: .8480481

**c.** tan 19° = .3443 by table; by calculator, press [19] [tan]: 0.3443276 □

Notice that Table II rounds the ratios to four decimal places, and that your calculator rounds the ratios at 8, 10, or 12 decimal places. If you round each calculator answer to four decimal places, it should be the same as the corresponding value in Table II.

Trigonometric ratios are useful in a variety of situations, as illustrated in the following example.

**EXAMPLE 17** The angle from the ground to the top of the Great Pyramid of Cheops is 52° if the distance from a point on the ground directly below the top is 351 ft. What is the height of the pyramid?

*Solution* From the figure at right, you can see that the height of the pyramid ($h$) is the side opposite the 52° angle; the distance 351 ft is the length of the adjacent side. Use a tangent ratio:

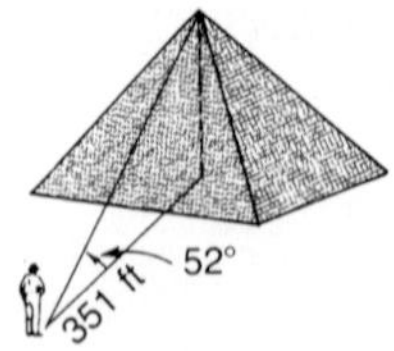


$$\tan 52^\circ = \frac{h}{351}$$

Solve for $h$ by multiplying both sides by 351:

$$h = 351 \tan 52^\circ$$

Now we must find the ratio for tan 52°. By Table II, tan 52°= 1.2799, so

$$\begin{aligned} h &= 351(1.2799) \\ &= 449.2449 \end{aligned}$$

By calculator, press [351] [×] [52] [tan] [=]: 449.25951.

The height given in our answer should not be more accurate than the given measurements, so we round this answer (either table or calculator) to the nearest foot. The height is 449 ft. □

Notice that the calculator and table answers in Example 17 are not identical. This is because both the table and the calculator give approximations of the exact value of tan 52°.

## 5.5 PROBLEM SET 5

**A**

1. In $\triangle TRI$ and $\triangle ANG$ shown at right, $\angle R \simeq \angle N$ and $TR = AN$. Name other pairs of parts you would need to know are equal in order to show that the triangles are congruent by
   **a.** SSS **b.** SAS **c.** ASA

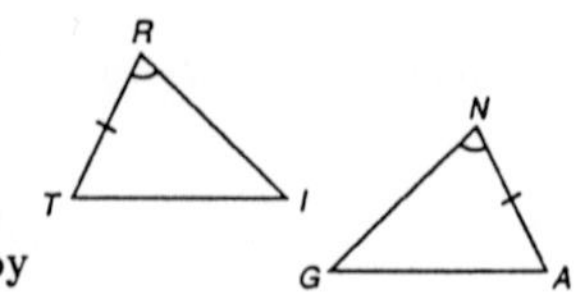


2. In $\triangle ABC$ and $\triangle DEF$ shown at right, $\angle A \simeq \angle D$ and $AC = DF$. Name other pairs of parts you would need to know are equal in order to show that the triangles are congruent by
   **a.** SSS **b.** SAS **c.** ASA

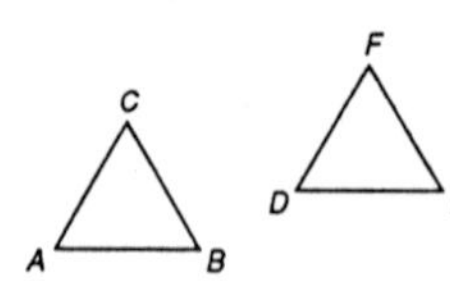


*Name the corresponding parts of the triangles in Problems 3–6.*

**3.**

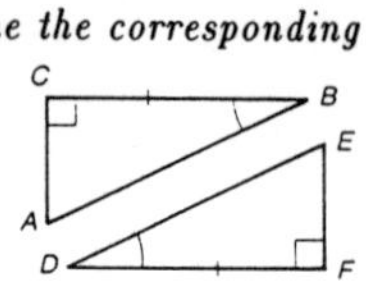


**4.**

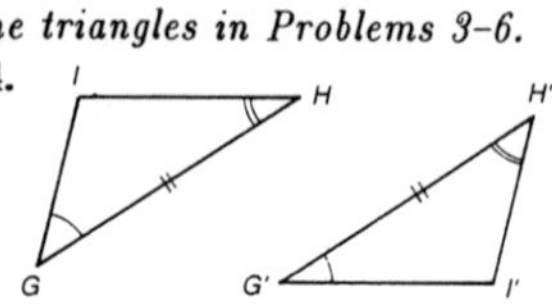

**5.**

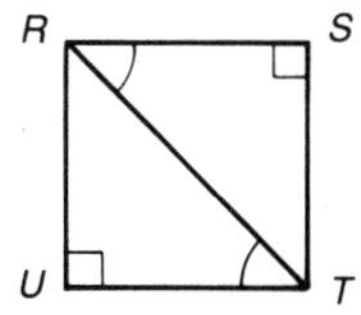


**6.**

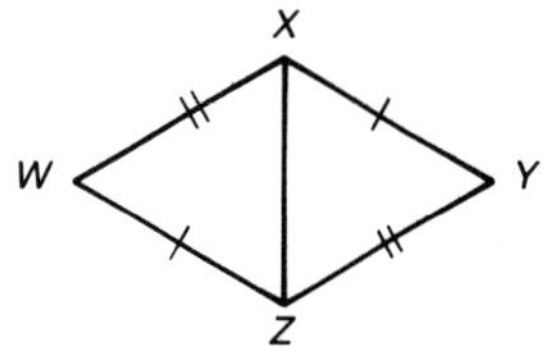


**7.** If $\triangle ABC \simeq \triangle DEF$, name the six corresponding parts.

**8.** If $\triangle REC \simeq \triangle TAN$, name the six corresponding parts.

**9.** If rectangle $DEFG$ is similar to rectangle $DABC$, and $DE = 4$, $DA = 24$, and $AB = 12$, what is the length of $EF$?

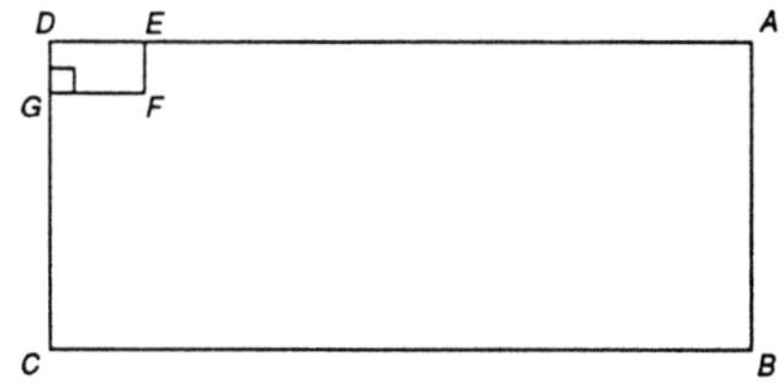


**10.** If polygon $HIJKLM$ is similar to polygon $ABCDEF$, and $HI = 10$, $AB = 5$, and $m\angle K = 40°$, what is $m\angle D$?

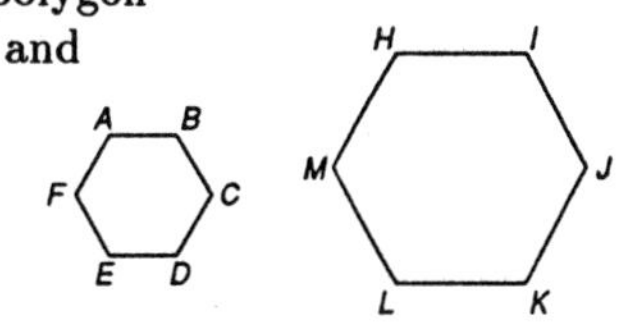


*In Problems 11–14, determine if each pair of triangles is congruent. If so, cite one of the congruent triangle properties.*

**11.**

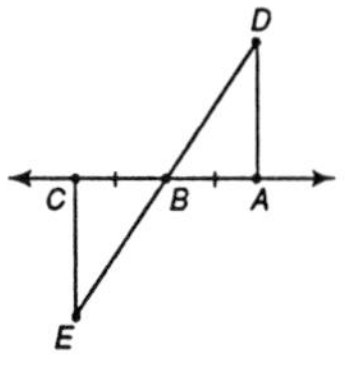


**12.**

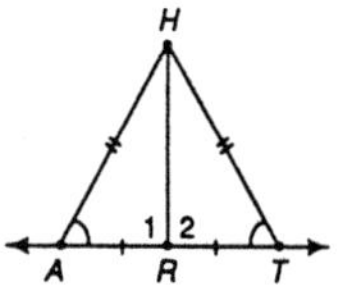


**13.**

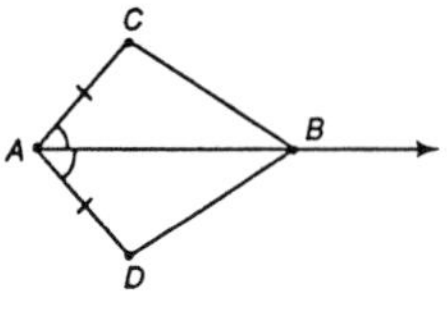


**14.**

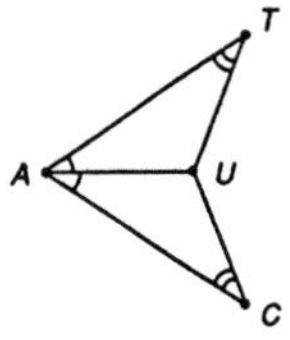


*In Problems 15–20, tell whether it is possible to conclude that the pairs of triangles are similar. If they are similar, list the proportional parts.*

**15.**

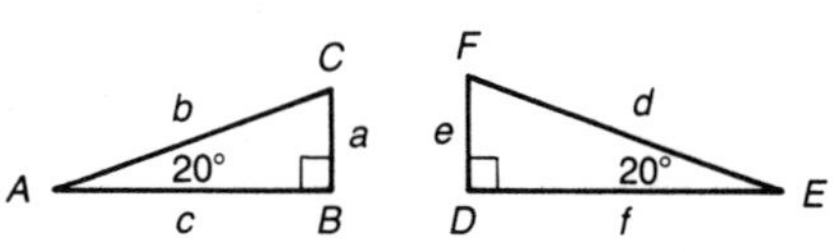

**16.**

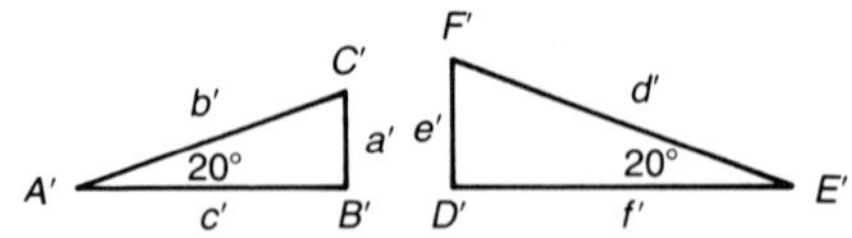


**17.**

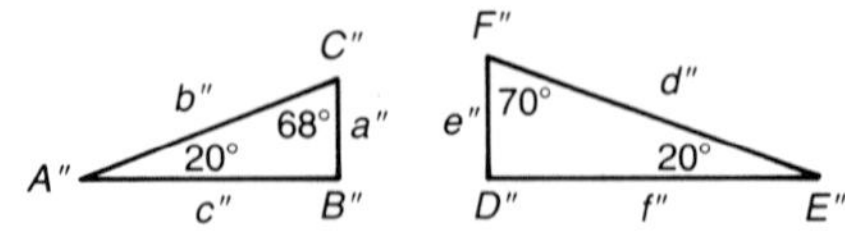


**18.**

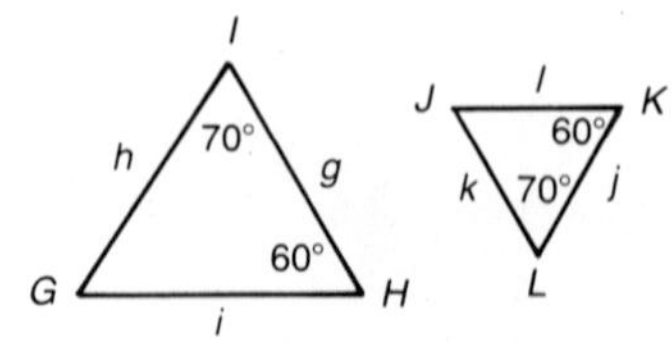


**19.**

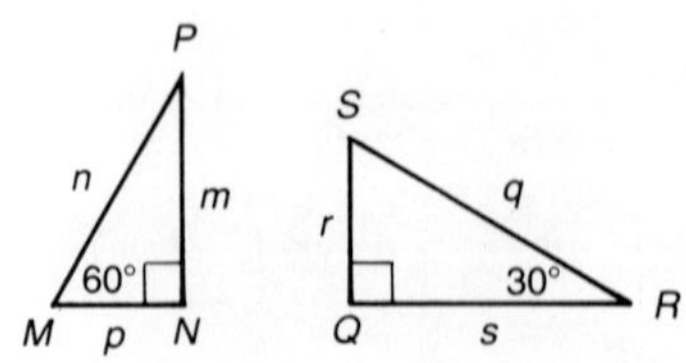


**20.**

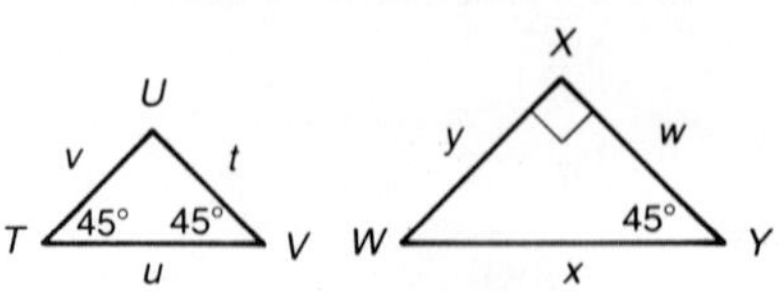


## B

*Given the two similar triangles shown at right, find the unknown lengths in Problems 21–26.*

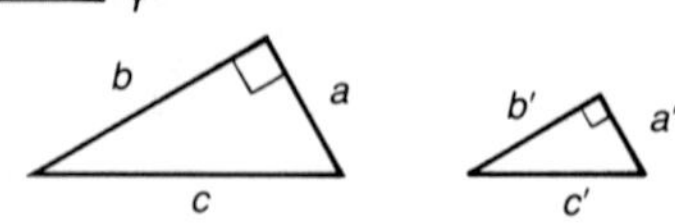


**21.** $a = 4$, $b = 8$, $a' = 2$; find $b'$.
**22.** $b = 5$, $c = 15$, $b' = 2$; find $c'$.
**23.** $c = 6$, $a = 4$, $c' = 8$; find $a'$.
**24.** $a' = 7$, $b' = 8$, $a = 5$; find $b$.
**25.** $b' = 8$, $c' = 12$, $c = 4$; find $b$.
**26.** $c' = 9$, $a' = 2$, $c = 5$; find $a$.

*Find the indicated lengths in Problems 27–36, using the following figure.*

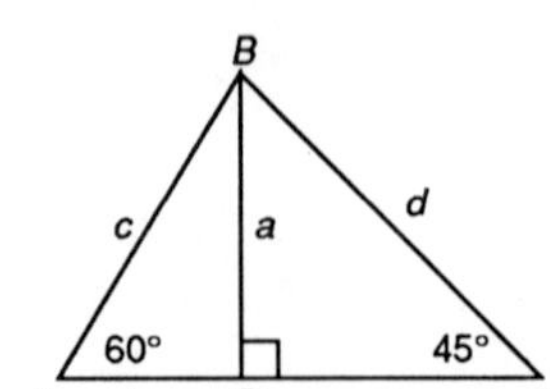


**27.** Given $x = 5$, find $d$.
**28.** Given $b = 8$, find $a$.
**29.** Given $x = 6$, find $a$.
**30.** Given $b = 10$, find $c$.
**31.** Given $d = 7$, find $a$.
**32.** Given $c = 15$, find $b$.
**33.** Given $d = 9$, find $x$.
**34.** Given $c = 12$, find $a$.
**35.** Given $x = 10$, find $c$.
**36.** Given $b = 14$, find $d$.

*Find the trigonometric ratios in Problems 37–48 by using either Table*

*II or a calculator.*

| | | | |
|---|---|---|---|
| **37.** sin 56° | **38.** sin 15° | **39.** sin 61° | **40.** sin 18° |
| **41.** cos 54° | **42.** cos 8° | **43.** cos 90° | **44.** cos 34° |
| **45.** tan 24° | **46.** tan 52° | **47.** tan 75° | **48.** tan 89° |

*In Problems 49–57, find the sine, cosine, and tangent for the angle A.*

**49.**

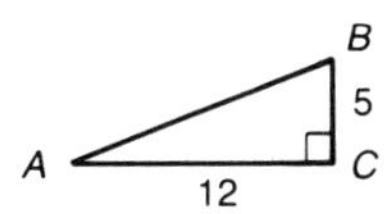


**50.**

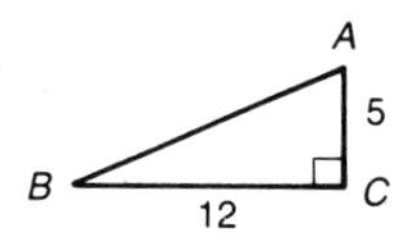


**51.**

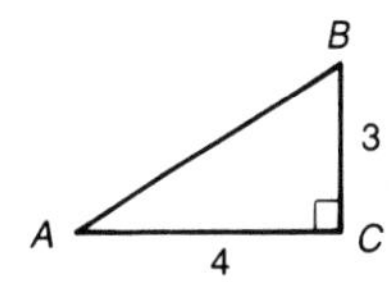


**52.**

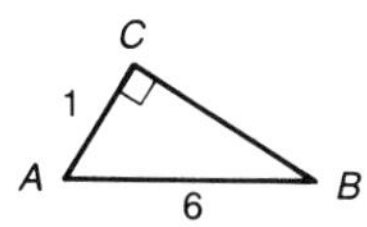


**53.**

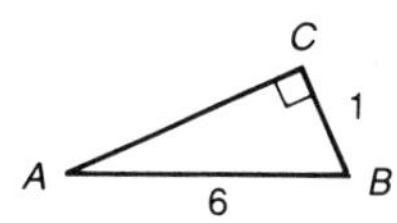


**54.**

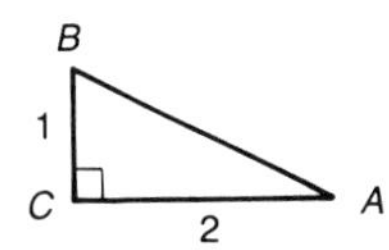


**55.**

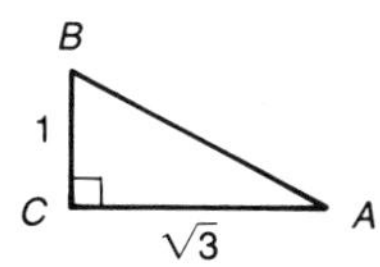


**56.**

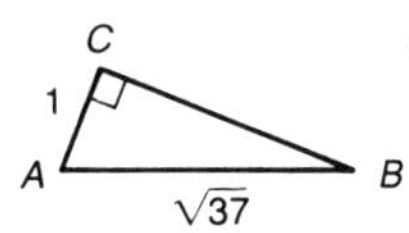


**57.**

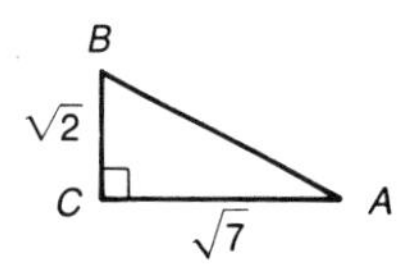


**58.** How far from the base of a building must a 26-ft ladder be placed so that it reaches 10 ft up the wall?

**59.** How high up a wall does a 26-ft ladder reach if the bottom of the ladder is placed 6 ft from the building?

**60.** A carpenter wants to be sure that the corner of a building is square and measures 6 ft and 8 ft along the sides. How long should the diagonal be?

**61.** What is the exact length of the hypotenuse if the legs of a right triangle are 2 in. each?

**62.** What is the exact length of the hypotenuse if the legs of a right triangle are 3 ft each?

**63.** An empty lot is 400 ft by 300 ft. How many feet would you save by walking diagonally across the lot instead of walking the length and width?

**64.** A television antenna is to be erected and held by guy wires. If the guy wires are 15 ft from the base of the antenna and the antenna is 10 ft high, what is the exact length of each guy wire? What is the length of each guy wire rounded to the nearest foot? If three guy wires are attached, how many feet of wire should be purchased, if it can not be bought in fractions of a foot?

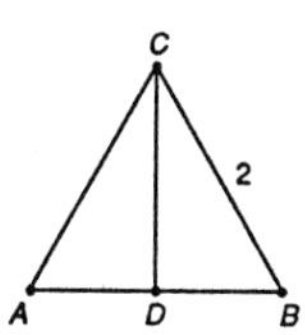


**65.** In equilateral $\triangle ABC$, shown in the margin, $D$ is the midpoint of segment $AB$. What is the length of $CD$?

**66.** In the figure shown at right, $AB$ and $DE$ are parallel. What is the length of $AB$?

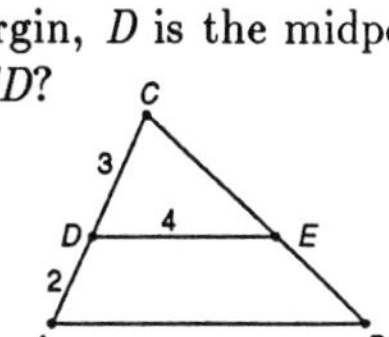

**67.** Carol walked diagonally across a rectangular field that measures 100 ft by 240 ft. How far did Carol walk?

**68.** Use similar triangles and a proportion to find the length of the lake shown in the figure below.

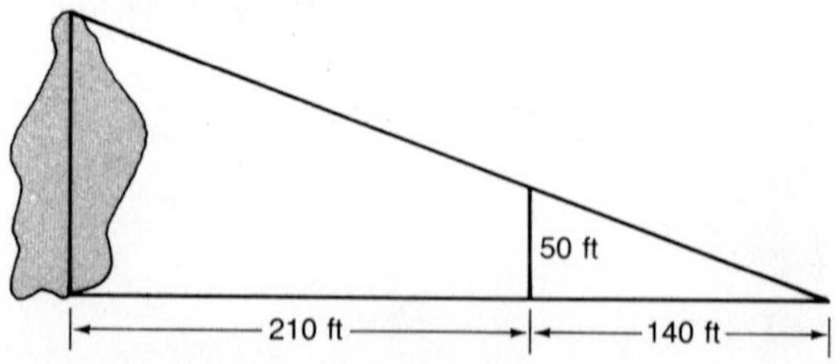


**69.** Use similar triangles and a proportion to find the height of the house shown in the figure below.

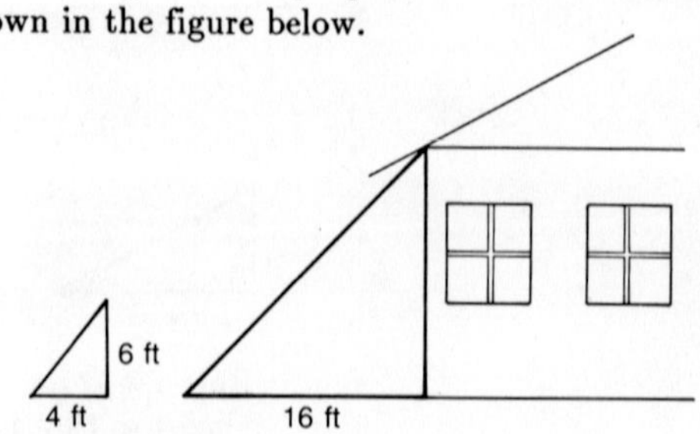


**70.** If a tree casts a shadow of 12 ft at the same time a 6-ft person casts a shadow of $2\frac{1}{2}$ ft, find the height of the tree to the nearest foot.

**71.** If the angle from the horizontal to the top of a building is 38° and the horizontal distance from its base is 90 ft, what is the height of the building (to the nearest foot)?

**72.** If the angle from the horizontal to the top of a tower is 52° and the horizontal distance from its base is 85 ft, what is the height of the tower (rounded to the nearest foot)?

# *CHAPTER 6 INDUCTIVE AND DEDUCTIVE REASONING

Two types of reasoning are commonly used in mathematics: inductive reasoning and deductive reasoning.

### 6.1 Inductive Reasoning

Observing patterns and then predicting answers or drawing conclusions based on observations is called **inductive reasoning**, or the **scientific method**. It involves reasoning from particular facts or individual cases to a general **conjecture** – a statement you think may be true. Thus, a generalization is made on the basis of some observed occurrences.

INDUCTIVE REASONING

> **Inductive reasoning** involves reaching a conclusion after making particular observations.

The more individual occurrences you observe, the better able you are to make a correct generalization. You must keep in mind that an inductive conclusion is always tentative and may have to be revised on the basis of new evidence. For example, consider this pattern:

$$\begin{aligned} 1 \times 9 &= 9 \\ 2 \times 9 &= 18 \\ 3 \times 9 &= 27 \\ 4 \times 9 &= 36 \\ 5 \times 9 &= 45 \\ 6 \times 9 &= 54 \\ 7 \times 9 &= 63 \\ 8 \times 9 &= 72 \\ 9 \times 9 &= 81 \\ 10 \times 9 &= 90 \end{aligned}$$

You might have made the conjecture that the sum of the digits of the answer to any product involving 9 is always 9:

$$\begin{aligned} 1 + 8 &= 9 \\ 2 + 7 &= 9 \\ &\vdots \end{aligned}$$

*Optional chapter

Certainly the first ten examples substantiate this speculation. But the conjecture is suspect because it is based on so few cases. It is shattered by the very next case, where we get $11 \times 9 = 99$, since $9 + 9 \neq 9$.

On the other hand, we can make a prediction about the exact time of sunrise and sunset tomorrow. This is also an example of inductive reasoning, since our prediction is based on a large number of observed cases. Thus, there is a very high probability that we will be successful in our prediction about the time of sunrise tomorrow.

**EXAMPLE 1**

Use inductive reasoning to find the next number in the sequences.

**a.** 6, 9, 12, 15, ...
**b.** 3, 12, 48, 192, ...
**c.** 4, 11, 15, 26, ...
**d.** 113, 322, 531, 740, ...
**e.** 1, 3, 2, 4, 3, 5, 4, ...

*Solution*

**a.** This sequence has what is called a *common difference,* which is found by subtracting successive terms of the sequence. In this example it is 3; so to find the next term, add this common difference to the last term:

$$15 + 3 = 18$$

**b.** This sequence has what is called a *common ratio*, which is found by dividing successive terms of the sequence. In this example it is 4; so to find the next term, multiply the last term by this common ratio:

$$192(4) = 768$$

**c.** This sequence is generated by added successive terms:

$$4 + 11 = 15$$
$$11 + 15 = 26$$

Thus, the next term is found by adding

$$15 + 26 = 41$$

**d.** This sequence has a common difference of 209, so add this to the last term:

$$740 + 209 = 949$$

**e.** This sequence exhibits an alternating pattern. In an alternating pattern, you use a different pattern rule on alternating terms. In this example, add 2 to the first number to get the second number. Then, subtract 1 from the second number to get the third number. To continue this sequence, alternately add 2 to one number and then subtract 1 from the next number. The next term is

$$4 + 2 = 6$$ □

Patterns need not involve only numbers. They may involve letters or geometric figures.

**EXAMPLE 2**

What is the next term in the sequence *A, C, F, J, O,* ... ?

*Solution*

After the *A,* one letter is omitted; then *C*; then two letters are omitted; then $F$; then three letters omitted; then $J$; then four letters omitted; then $O$; so the next gap should have five letters omitted; thus, the next letter in the pattern is $U$. □

**EXAMPLE 3** What is the next geometric shape in the following sequence?

The pattern rotates the triangle 90° counterclockwise to create the next term in the sequence. Thus, the next shape is

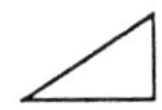

□

### 6.2 Deductive Reasoning

**Deductive reasoning** produces results that are *certain* within the logical system being developed. For example, consider the following argument:

1. If you take the *Times*, then you are well-informed.
2. You take the *Times*.
3. Therefore, you are well-informed.

If you accept statements 1 and 2 are true, then you *must* accept statement 3 as true. Statements 1 and 2 are called the **premises** of the argument, and statement 3 is called the **conclusion.** Such reasoning is called **deductive reasoning,** and if the conclusion follows from the premises, the reasoning is said to be **valid.**

DEDUCTIVE REASONING

> **Deductive reasoning** involves reaching a conclusion by using a formal structure based on a set of **undefined terms** and a set of accepted unproved **axioms or premises.** The conclusions are said to be **proved** and are called **theorems.**

**Logic** began to flourish during the classical Greek period. Aristotle (384–322 B.C.) was the first person to study the subject systematically, and he and many other Greeks searched for universal truths that were irrefutable. The logic of this period, referred to as **Aristotelian logic,** is still used today. The example of deductive reasoning given above is an example of Aristotelian logic.

The second great period for logic came with the use of symbols to simplify complicated logical arguments. This began when the great German mathematician Gottfried von Leibniz (1646–1716), at the age of 14, attempted to reform Aristotelian logic. He called his logic the **universal characteristic** and wrote, in 1666, that he wanted to create a general method in which truths of reason would be reduced to a calculation so that errors of thought would appear as computational errors. You will study more about this if you take a course in symbolic logic.

However, the world took little notice of Leibniz's logic, and it wasn't until George Boole (1815–1864) completed his book, *An Investigation of the Laws of Thought*, that deductive logic entered its third and most important period. Boole considered various mathematical operations by separating them from other commonly used symbols. This idea was popularized by Bertrand Russell (1872–1970) and Alfred North Whitehead (1861–1947) in their monumental *Principia Mathematica.* In this work, they began with a few assumptions and three undefined terms and built a system of symbolic logic. From this, they formally developed many theorems of arithmetic and mathematics.

Logic accepts no conclusions except those that are inescapable. This is possible because of the strict way in which concepts are defined. Difficulty in simplifying arguments may arise because of their length, the vagueness of the words used, their literary style, or the possible emotional impact of the words used. Consider the following two arguments:

1. If George Washington was assassinated, then he is dead.
   Therefore, if he is dead, he was assassinated.

2. If you use heroin, then you first used marijuana.
   Therefore, if you use marijuana, then you will use heroin.

Logically these two arguments are exactly the same, and both are **invalid** forms of reasoning. Nearly everyone would agree that the first is invalid, but many people see the second as valid. The reason here is the emotional appeal of the words used.

To avoid these difficulties, we look at the *form* of the arguments and not the truth or falsity of the statements. For the rest of this section, we'll focus on one type of logic problem called a **syllogism.** A syllogism has three parts: two *premises,* or hypotheses, and a *conclusion.* The premises give us information from which we form a conclusion. With the syllogism we are interested in knowing whether the conclusion *necessarily follows* from the premises. If it does, it is called a **valid syllogism**; if not, it is called **invalid.**

**EXAMPLE 4** Valid forms of reasoning:

All Chevrolets are automobiles.
All automobiles have four wheels.
Therefore, all Chevrolets have four wheels.

All teachers are crazy.
Karl Smith is a teacher.
Therefore, Karl Smith is crazy. □

**EXAMPLE 5** Invalid forms of reasoning:

Some people are nice people.
Some nice people are broke.
Therefore, some people are broke.

All dodos are extinct.
No dinosaurs are dodos.
Therefore, all dinosaurs are extinct. □

In order to analyze such arguments, we need to have a systematic method of approach. We will use something called a **Venn diagram.** For two statements $p$ and $q$, draw two intersecting circles. The premises may be of the form:

Some $p$ are $q$.
No $p$ are $q$.
All $p$ are $q$.

We can illustrate these argument forms as shown in Figure 32. Place an $x$ in a region *known* to be occupied (as with the statement, "Karl Smith is a teacher") and shade in a region known to be empty (as with the statement "No dinosaurs are dodos"). Sometimes it is convenient to mark an empty region with $\emptyset$ instead of shading it. For the statement, "All $p$ are $q$," the region called $p$ that is not contained in $q$ is empty.

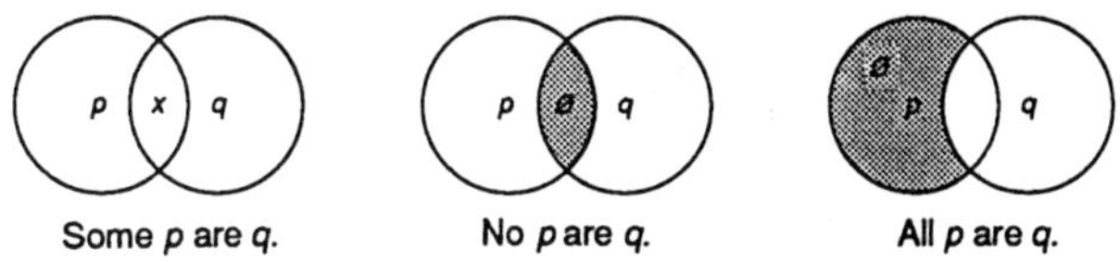

Figure 32 Venn diagrams

The Venn diagram for three propositions is shown in Figure 33.

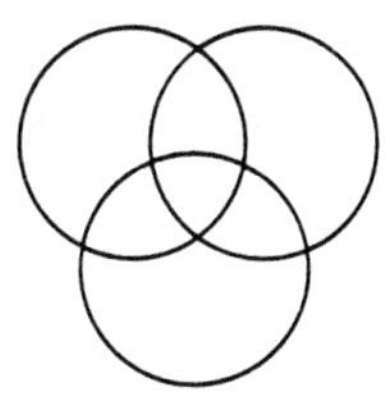

Figure 33 Venn diagram for three propositions

**EXAMPLE 6**

If the regions of a Venn diagram are labeled as shown, describe each numbered region in words.

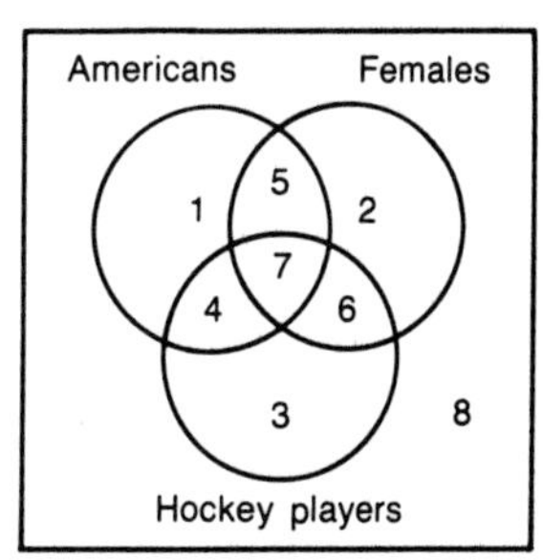


*Solution*

**Region 1:** Male Americans who are not hockey players
**Region 2:** Non-American females who are not hockey players
**Region 3:** Non-American male hockey players
**Region 4:** American male hockey players
**Region 5:** Female Americans who are not hockey players
**Region 6:** Non-American female hockey players
**Region 7:** American female hockey players
**Region 8:** Non-American males who are not hockey players □

**EXAMPLE 7**

Test the validity of the following argument.

All dictionaries are useful.
All useful books are valuable.
Therefore, all dictionaries are valuable.

*Solution* Let $D$: dictionaries; $U$: useful objects; and $V$: valuable objects. Draw a Venn diagram.

First premise:

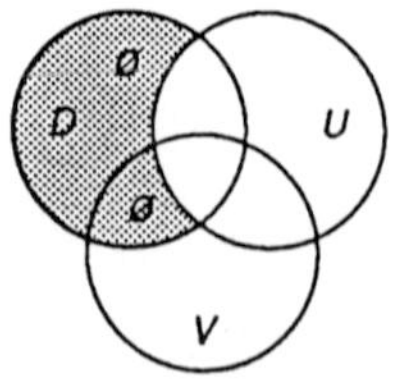


Second premise:

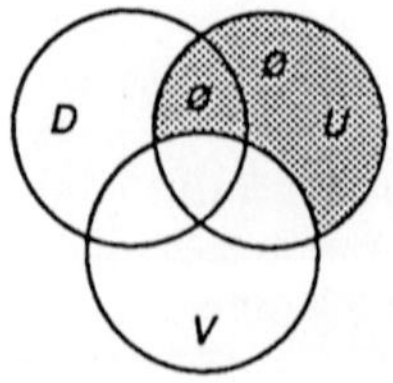


Both premises together:
(Your work will look like this.)

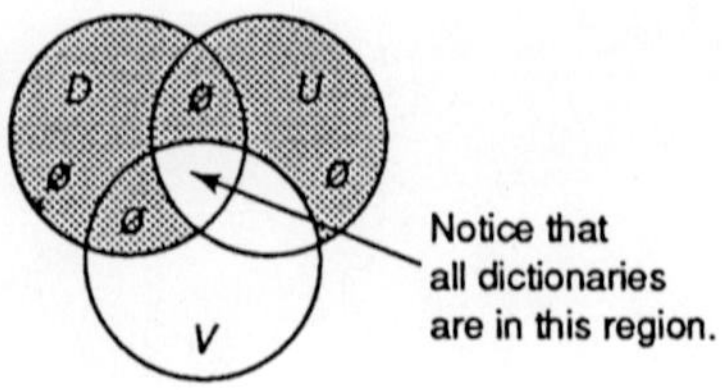


The conclusion is valid. □

**EXAMPLE 8** Test the validity of the following argument.

If you like potato chips, then you will like Krinkles.
You do not like potato chips.
Therefore, you do not like Krinkles.

*Solution* Let $P$: people who like potato chips and $K$: people who like Krinkles.

Draw the Venn diagram.
Add $\emptyset$ from the first premise.
Premise 2 does not force $x$ into a single region; it could be in two places — labeled $x_1$ and $x_2$.

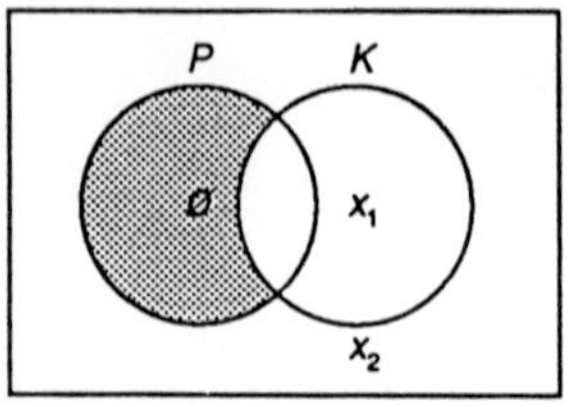


Since the conclusion is not forced, the argument is not valid. □

**6.3 Survey Problems**

Venn diagrams can sometimes be used for survey problems. Suppose a survey indicates that 45 students are taking mathematics and 41 are taking English. How many students are taking math or English? At first, it might seem that all we need to do is add 41 and 45, but such is not the case, as you can see by looking at Figure 34(a).

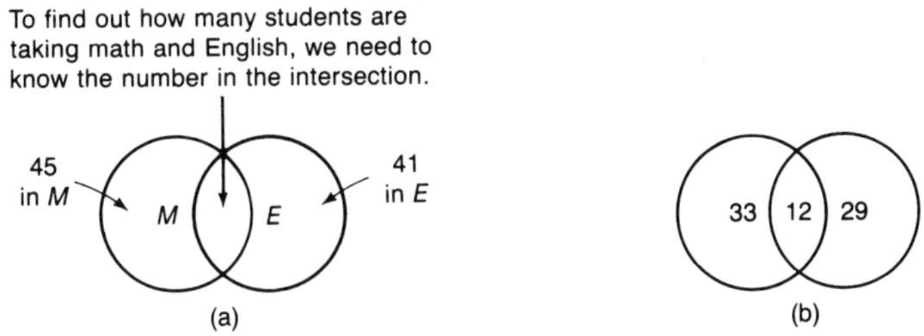


Figure 34

Suppose, further, that 12 students are taking both math and English. In this case, we can fill in the number in the intersection, as shown in Figure 34b. The number of students enrolled in mathematics or English is found by looking at Figure 34(b).

$33 + 12 + 29 = 74$ *These are the numbers in each region shown in Figure 34(b).*

For three sets, the situation is a little more involved. Remember, use deductive reasoning to fill in the number in the innermost set first, and then work your way outward through the Venn diagram by using subtraction.

EXAMPLE 9

A survey of 100 randomly selected students gave the following information:

45 are taking math;
41 are taking English;
40 are taking history;
15 are taking math and English;
18 are taking math and history;
17 are taking English and history;
7 are taking all three.

**a.** How many are taking only math?
**b.** How many are taking only English?
**c.** How many are taking only history?
**d.** How many are not taking any of these courses?
**e.** How many are not taking math?

*Solution*

Draw a Venn diagram and fill in the various regions. Let $M$: persons taking math; $E$: persons taking English; and $H$: persons taking history.

Fill in the innermost section first; there are 7 in this region.

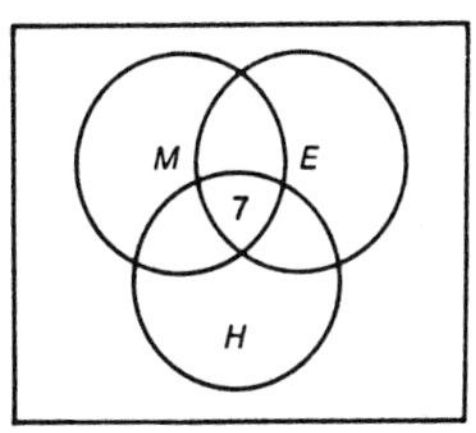


Step 1

Next, fill in the other inner regions:
fill in 10 (total 17),
fill in 11 (total 18), and
fill in 8 (total 15).

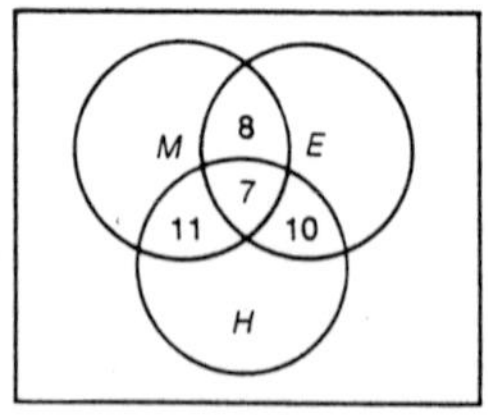


Step 2

Finally, fill in the remaining regions within each large circle:
fill in 12 in $H$ (total 40),
fill in 16 in $E$ (total 41), and
fill in 19 in $M$ (total 45).

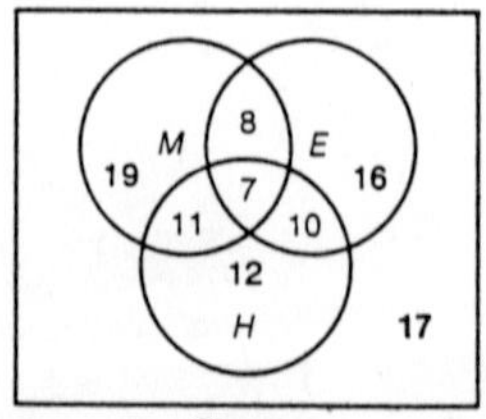


Step 3

There are 17 students not accounted for (83 have been listed; add them up) in all of the above regions. You can now use the final Venn diagram to answer the questions.

**a.** 19 **b.** 16 **c.** 12 **d.** 17 **e.** 55 □

## 6.4 PROBLEM SET 6

A

1. In your own words, explain inductive reasoning. Give an original example of when you have used inductive reasoning or heard it being used.
2. In your own words, explain deductive reasoning. Give an original example of when you have used deductive reasoning or heard it being used.
3. What is a syllogism? Distinguish between truth and validity.

*Use inductive reasoning to give the next two terms in each sequence in Problems 4–15.*

**4.** 2, 5, 8, 11, 14, ...
**5.** 1, 2, 1, 1, 2, 1, 1, 1, 2, 1, 1, 1, ...
**6.** 1, 2, 4, 7, 11, 16, ...
**7.** 1, 3, 4, 7, 11, 18, 29, ...
**8.** 3, 6, 12, 24, 48, ...
**9.** 5, 15, 45, 135, 405, ...
**10.** 100, 99, 97, 94, 90, ...
**11.** 1, 4, 9, 16, 25, ...
**12.** 2, 5, 2, 5, 5, 2, 5, 5, 5, ...
**13.** 5, 15, 25, 35, ...
**14.** $A, B, B, E, C, I, D, \ldots$
**15.** $A, E, F, H, I, K, \ldots$
**16.** If the regions are labeled as shown below describe each region in words.

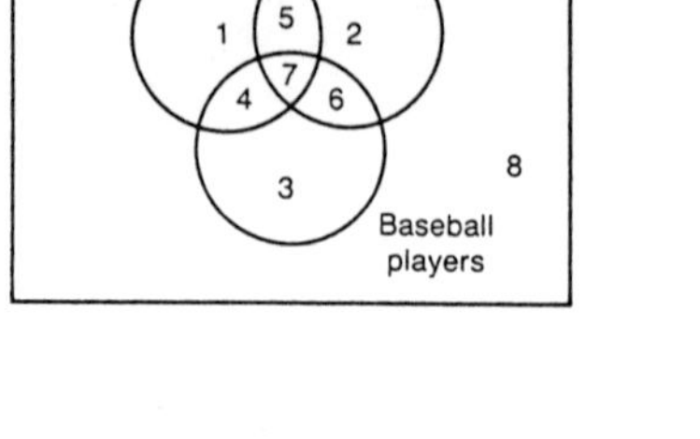

**17.** If the regions are labeled as shown below describe each region in words.

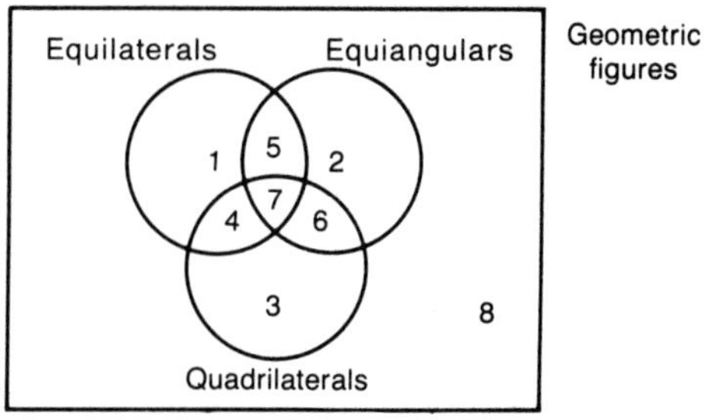


## B

*Use Venn diagrams to check the validity of the arguments in Problems 18–28.*

**18.** All mathematicians are eccentrics.
All eccentrics are rich.
Therefore, all mathematicians are rich.

**19.** Some beautiful women are blond.
Blonds have more fun.
Therefore, some beautiful women have more fun.

**20.** All bachelors are handsome.
Some bachelors do not drink lemonade.
Therefore, some handsome men do not drink lemonade.

**21.** No students are enthusiastic.
You are enthusiastic.
Therefore, you are not a student.

**22.** Some women are tall.
No men are strong.
Therefore, some tall people are not strong.

**23.** No politicians are honest.
Some dishonest people are found out.
Therefore, some politicians are found out.

**24.** All candy is fattening.
All candy is delicious.
Therefore, all fattening food is delicious.

**25.** No professors are ignorant.
All ignorant people are vain.
Therefore, no professors are vain.

**26.** No monkeys are soldiers.
All monkeys are mischievous.
Therefore, some mischievous creatures are not soldiers.

**27.** All lions are fierce.
Some lions do not drink coffee.
Therefore, some creatures that drink coffee are not fierce.

**28.** All red hair is pretty.
No pretty things are valuable.
Therefore, no red hair is valuable.

**29.** In a recent survey of 100 persons, the following information was gathered:

55 use shampoo $A$.
51 use shampoo $B$.
35 use shampoo $C$.
24 use shampoos $A$ and $B$.
19 use shampoos $A$ and $C$.
13 use shampoos $B$ and $C$.
11 use all three.

Let $A$: people who use shampoo $A$; $B$: people who use shampoo $B$; and $C$: people who use shampoo $C$. Use a Venn diagram to show how many people are in each of the eight possible categories.

**30.** Matt E. Matic is applying for a job. To see whether he could handle the job, the personnel manager sent him out to poll 100 people about their favorite types of TV shows. His data were as follows:

62 preferred comedies.
38 preferred variety shows.
42 preferred serious drama.
18 preferred comedies or variety shows.
12 preferred variety shows or serious drama.
16 preferred comedies or serious drama.
7 preferred all types.
2 didn't like any TV shows.

If you were the personnel manager, would you hire Matt on the basis of this survey?

CHAPTER 7

# SUMMARY AND REVIEW

In this chapter we look at ideas from geometry that you may need in your future work in mathematics. The material of this chapter is reviewed in the following list of objectives.

**7.1 Review**

1. Classify polygons according to the number of sides.
2. Classify angles as acute, right, straight, or obtuse.
3. Know relationships and terminology of angles formed by a transversal intersecting parallel lines.
4. Find the measure of the third angle of a triangle.
5. Find the measure of the exterior angle of a triangle.
6. Find the perimeter of a triangle, equilateral triangle, rectangle, and square.
7. Find the circumference of a circle or part of a circle.
8. Find the distance around a given figure.
9. Find the area of a rectangle, square, parallelogram, triangle, trapezoid, or combination of these.
10. Work problems involving unknown parts in the perimeter or area formulas.
11. Find the surface area of a solid.
12. Find the volume of a cube or rectangular solid.
13. Find the volume of a right prism, right circular cylinder, pyramid, right circular cone, or sphere.
14. Solve applied problems involving perimeter, area, and volume.
15. Name the corresponding parts of congruent triangles.
16. Determine whether two triangles are congruent.
17. Determine whether two triangles are similar.
18. Use proportions and similar triangles to find the length of a side of a triangle.
19. Know the Pythagorean Theorem.
20. Use the $45°$-$45°$-$90°$ property of triangles.
21. Use the $30°$-$60°$-$90°$ property of triangles
22. Find trigonometric ratios, using either a table or a calculator.
23. Find the sine, cosine, and tangent for an angle in a right triangle.
24. Use similar triangles, the Pythagorean Theorem, or trigonometric ratios to find an unknown distance.

*25. Distinguish between inductive and deductive reasoning.
*26. Use inductive reasoning to determine the next term in a pattern.
*27. Use deductive reasoning and Venn diagrams to check the validity of a given argument.
*28. Answer questions about a survey by using a Venn diagram.

## 7.2 PROBLEM SET 7

CONCEPT PROBLEMS

1. A polygon with four sides is called a(n) ____________.
2. An angle with measure less than 90° is classified as __________.
3. When parallel lines are cut by a transversal, the following pairs of angles are congruent: ____________ angles, __________ angles, __________ interior angles , and ____________ exterior angles.
4. Adjacent angles formed by two intersecting lines are __________ (i.e., angles whose measures add to 180°).
5. If two lines intersect to form congruent adjacent angles, the lines are ____________.
6. The sum of the measures of a triangle is ____________.
7. The formula for the perimeter of an equilateral triangle with side $s$ is ____________.
8. The formula for the perimeter of a rectangle with length $l$ and width $w$ is ____________.
9. The formula for the circumference of a circle with diameter $d$ is ____________.
10. In a circle the relationship between the diameter and the radius is ____________.
11. The formula for the area of a rectangle with length $l$ and width $w$ is ____________.
12. The formula for the area of a square with side $s$ is __________.
13. An acre is ____________ $ft^2$.
14. The formula for the area of a parallelogram with base $b$ and height $h$ is ____________.
15. The formula for the area of a trapezoid with bases $b$ and $B$ and height $h$ is ____________.
16. The formula for the area of a circle with radius $r$ is __________.
17. To find the surface area of a solid, ____________ the area of the ____________.
18. The formula for the volume of a cube with edge $s$ is __________.
19. The formula for the volume of a rectangular solid with edges $l$, $w$, and $h$ is ____________.
20. The formula for the volume of a right prism or a right circular cylinder whose base has area $B$ and whose height is $h$ is ______.
21. The formula for the volume of a sphere is ____________.
22. The formula for the volume of a pyramid or a right circular cone whose base has area $B$ and whose height is $h$ is __________.
23. To convert cubic feet into cubic yards ____________.
24. Two triangles are ____________ if their corresponding sides have the same length and the corresponding angles have the same measure.

*Optional chapter

25. The three congruent triangle properties are called ________ .
26. Two triangles are similar if two angles of one triangle have the same measure as two angles of the other triangle. If the triangles are similar, then ________ .
27. The Pythagorean Theorem states ________ .
28. In a right triangle $ABC$ with right angle at $C$, $\sin B =$ ____ , $\cos B =$ ________ , and $\tan B =$ ________ .
*29. Inductive reasoning is reaching a conclusion by ________ .
*30. Deductive reasoning is reaching a conclusion by ________ .
*31. A syllogism has two or more ________ and a(n) ________ .

PRACTICE PROBLEMS

32. Give the name of the polygon with the given number of sides.
**a.** six **b.** four **c.** eight **d.** three **e.** five

33. Give the name of each polygon.
**a.**
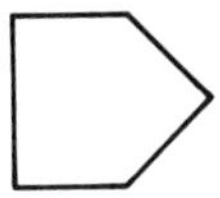
**b.**
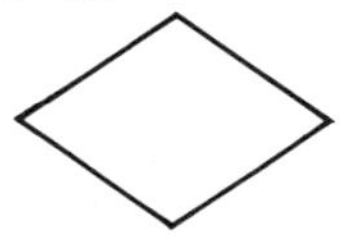
**c.**
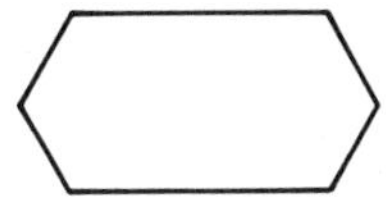

34. In Figure 35, classify the given angle as acute, right, straight, or obtuse.
**a.** $\angle BOS$ **b.** $\angle ESO$ **c.** $\angle ROS$ **d.** $\angle BTL$ **e.** $\angle POS$

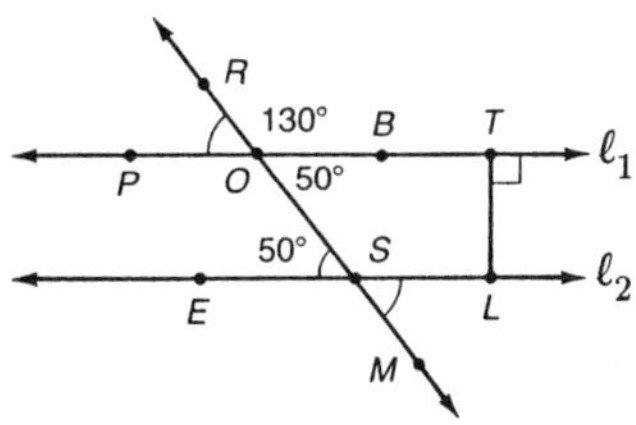


Figure 35

35. In Figure 35, assume that $\ell_1$ and $\ell_2$ are parallel. Find the measures of the given angles, and give a reason for your answer.
**a.** $\angle LSM$ **b.** $\angle POS$ **c.** $\angle TLS$ **d.** $\angle ROP$ **e.** $\angle ESM$

36. In $\triangle ABC$, what is the measure of angle $A$ if $m\angle B = 58^\circ$ and $m\angle C = 37^\circ$?

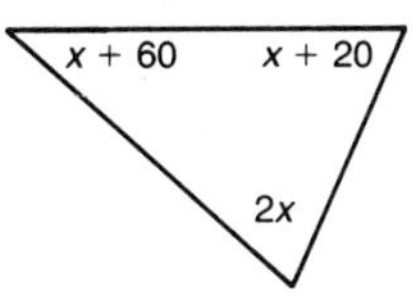


37. What are the measures of the angles in the figure in the margin?
38. Find the value of $x$ in each of the figures.
**a.**
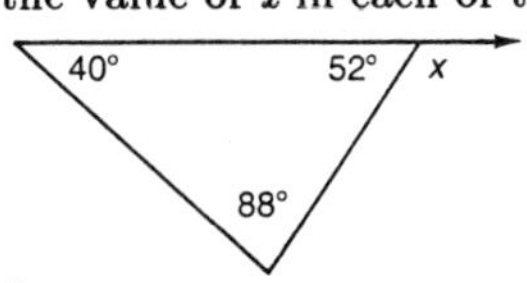

**b.**
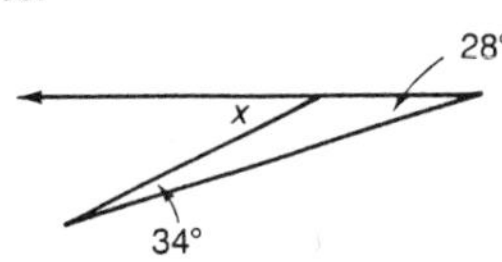

**c.**
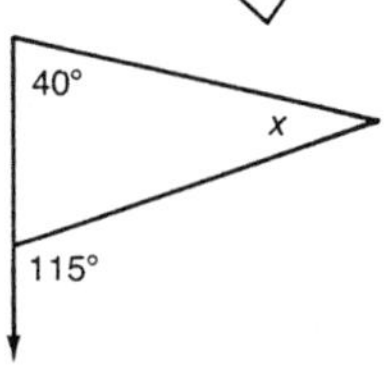

**d.**
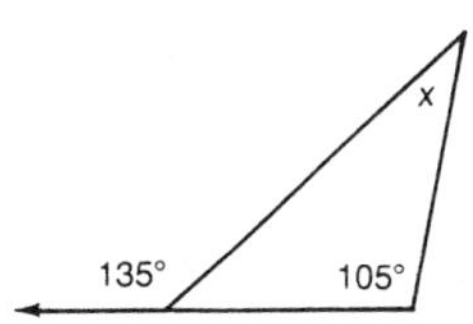


*Optional chapter

*Find the perimeters in Problems 39–42.*

**39.**

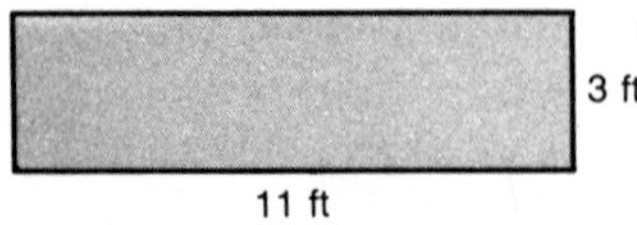


**40.**

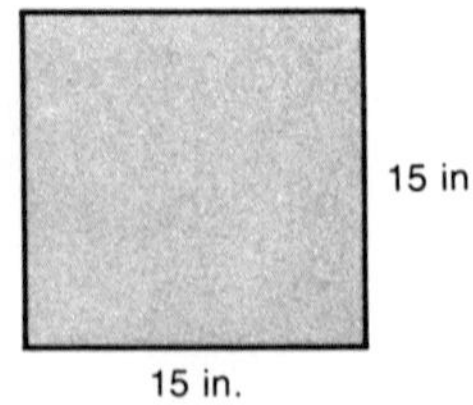


**41.**

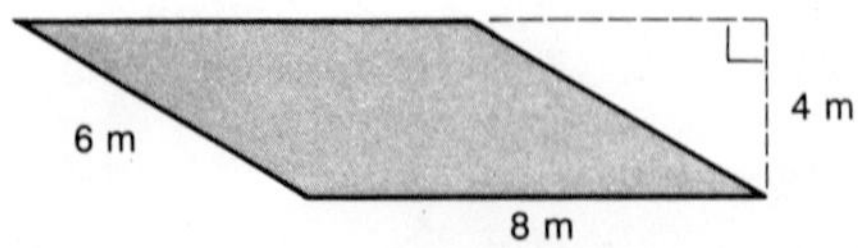


**42.**

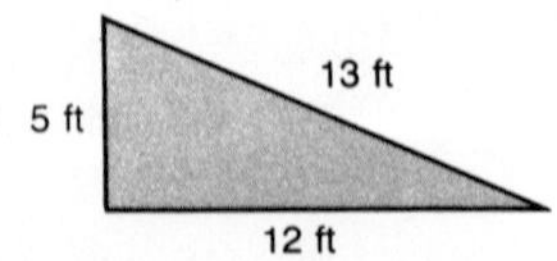


*Find the circumference or the distance around the shaded portions in each of Problems 43–46. Round your answer to two decimal places.*

**43.**

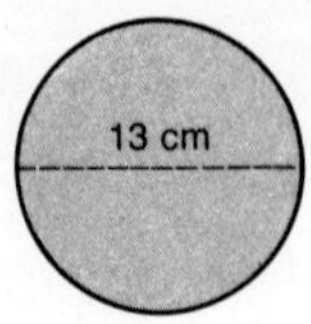


**44.**

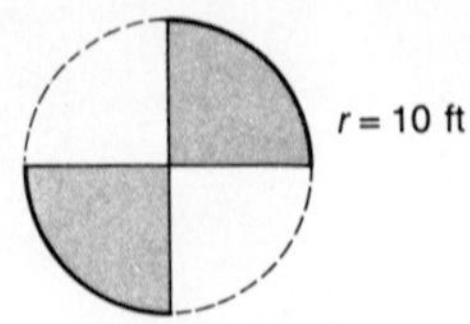


**45.**

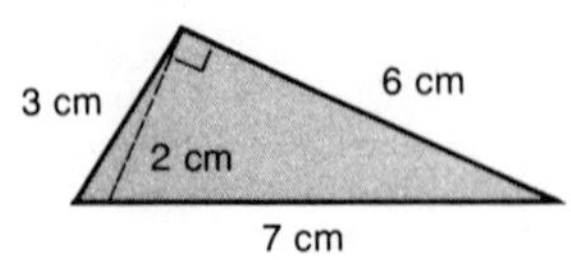


**46.**

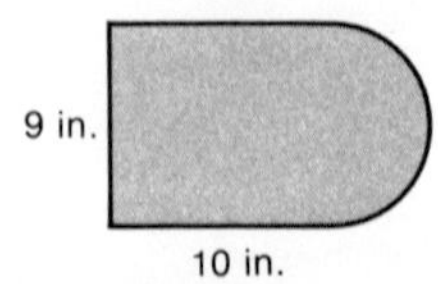


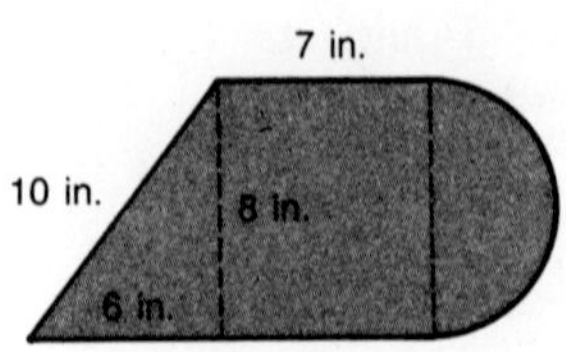


**47.** Find the area of the shaded portion of the figure in Problem 43.
**48.** Find the area of the shaded portion of the figure in Problem 44.
**49.** Find the area of the shaded portion of the figure in Problem 45.
**50.** Find the area of the shaded portion of the figure in Problem 46.
**51.** Find the distance around and the area of the figure in the margin. Round your answer to the nearest unit.
**52.** What is the area of a circle whose circumference is $10\pi$?
**53.** If the perimeter of a rectangle is 15 in. and one side has length 3 in., what is the area of this rectangle?
**54.** If the area of a square is 9 $cm^2$, what is the perimeter?

*Find the surface areas (rounded to the nearest unit) of the solids shown in Problems 55–58.*

**55.**

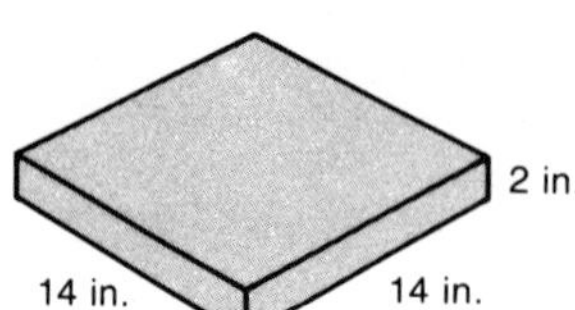


**56.**

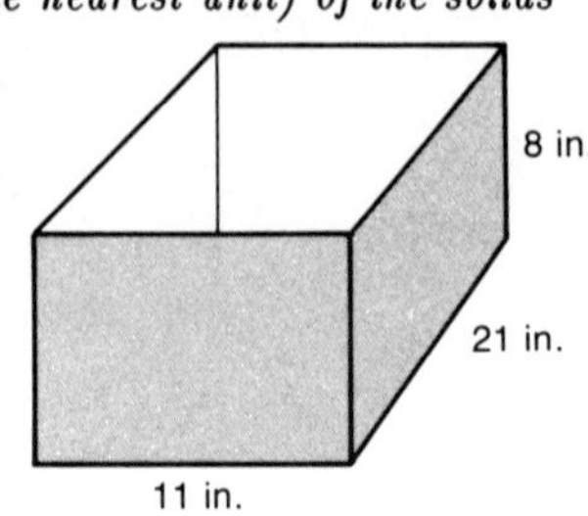

**57.**

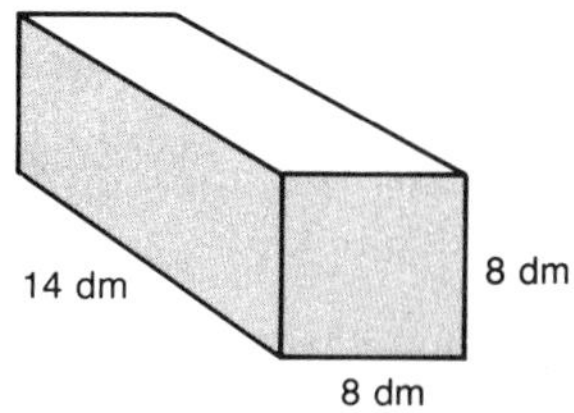


**58.**

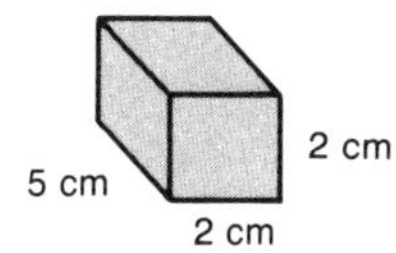


**59.** Find the volume of the solid in Problem 55.
**60.** Find the volume of the solid in Problem 56.
**61.** Find the volume of the solid in Problem 57.
**62.** Find the volume of the solid in Problem 58.

*Find the volumes (rounded to the nearest cubic unit) in Problems 63–66.*

**63.**

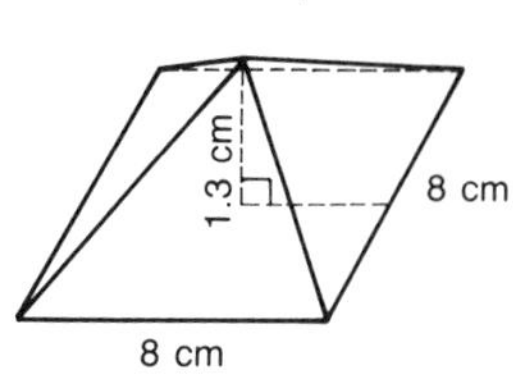


**64.**

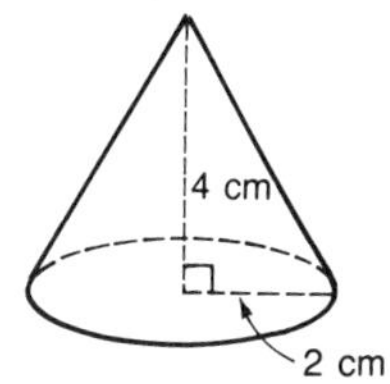


**65.**

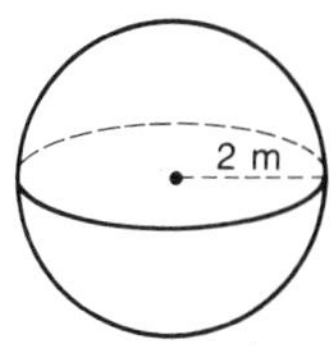


**66.**

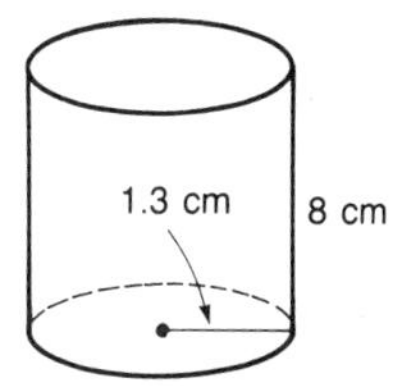


**67.** What is the volume (to the nearest tenth of a cubic foot) of a sphere with radius 1 ft?

**68.** How many square yards of carpet are necessary to carpet a 13-ft by 14-ft room? (Assume that you cannot purchase part of a square yard.)

**69.** If you are charged $100 per square foot for custom stained glass windows, how much should you pay for a 2-ft by 2-ft window? How about a circular window with a 2-ft diameter?

**70.** How many cubic yards of concrete are necessary for a rectangular driveway that is 20 ft by 25 ft by 4 in.? Assume that you can order in units of $\frac{1}{2}$ cubic yard.

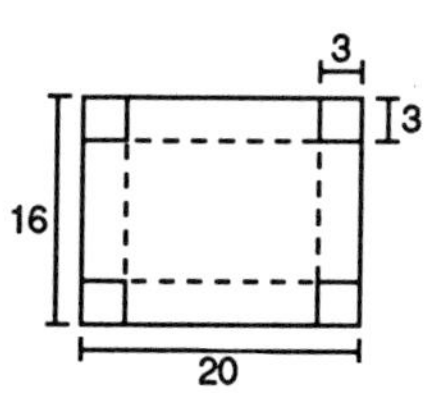


**71.** A 3-in. square is cut from each corner of a 20-in. × 16-in. sheet of metal. The metal is folded along the dashed lines, as shown in the diagram in the margin. What is the volume of the pan thus formed?

**72.** A pedestal for a statue is to have a height of 4 ft and a diameter of 2 ft. The pedestal's base is a rectangular solid that is 8 ft by 5 ft by 2 ft, as shown in the diagram at right. What volume of cement (to the nearest one-half cubic yard) is needed to construct the pedestal and its base?

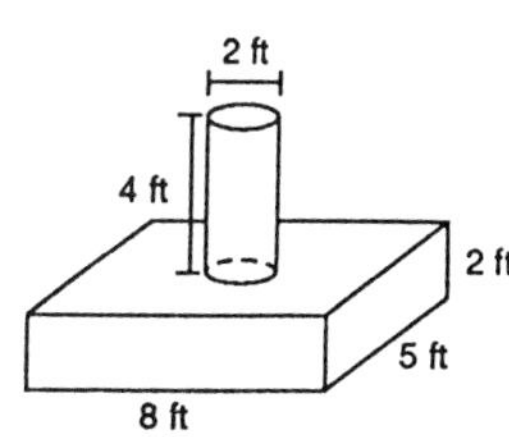


**73.** A right cylindrical tank with radius 5 ft and height 10 feet is full of water. If one-third of the water is drained from the tank, how many cubic feet (rounded to the nearest unit) of water remain in the tank?

74. If the edge of a cube is doubled in length, then the volume of the cube is increased by what factor?

75. If $\triangle TRI \simeq \triangle ARI$ in Figure 36, name the corresponding parts.

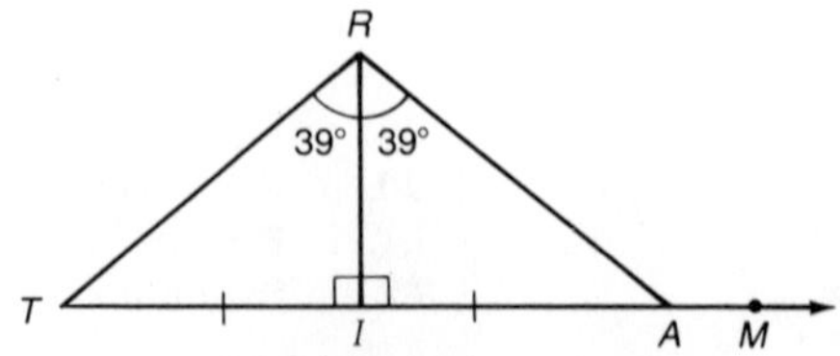


Figure 36

76. a. What is the measure of $\angle T$ in Figure 36?
b. What is the measure of $\angle RAM$ in Figure 36?
c. If $RI$ in Figure 36 is 6 in., what is the length of $AI$ (rounded to the nearest inch)?
d. Is $\triangle TRI \sim \triangle ARI$ in Figure 36? Why or why not?

77. If $\triangle ABC \sim \triangle A'B'C'$, $a' = 5$, $b' = 12$, and $a = 7\frac{1}{2}$, what is $b$?

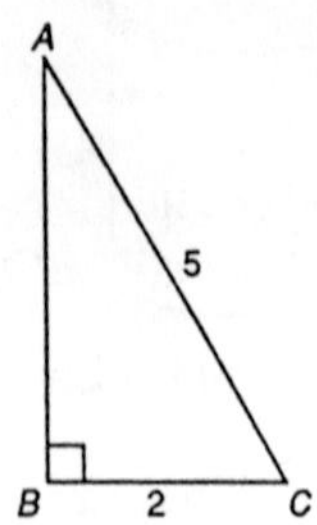


78. In the figure shown in the margin, what is the length of $AB$?

79. a. If $a = 4$, find $b$, using Figure 37.

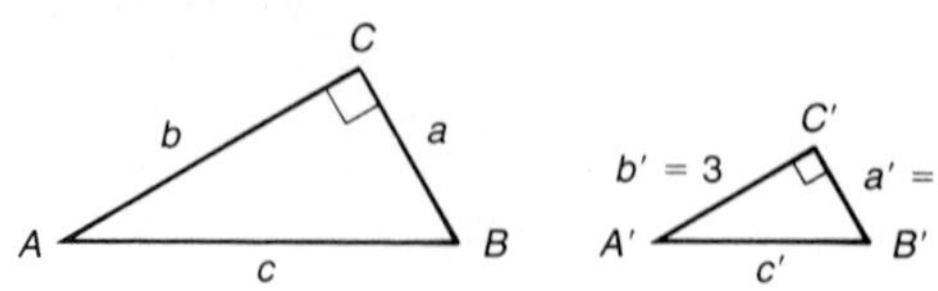


Figure 37

b. If $b = 4$, find $a$.
c. From the Pythagorean Theorem, find $c$ if $a = 4$.
d. From the Pythagorean Theorem, find $c$ if $b = 3.6$.

80. Using Figure 37,
a. Find $b$ if $m\angle A = 35°$ and $a = 45$.
b. Find $b$ if $m\angle B = 22°$ and $a = 38$.
c. Find $c$ if $m\angle B = 17°$ and $b = 19$.
d. Find $c$ if $m\angle B = 28°$ and $a = 128$.

81. Find the value of the given trigonometric ratio, using a table or a calculator.
a. sin 59° b. tan 0° c. cos 63° d. tan 82° e. sin 90°

82. Find the value of the given trigonometric ratio, using a table or a calculator.
a. tan 77° b. cos 83° c. cos 18° d. sin 69° e. sin 0°

83. In equilateral $\triangle ABC$ shown at right, $D$ is the midpoint of segment $AB$. What is the length of $CD$?

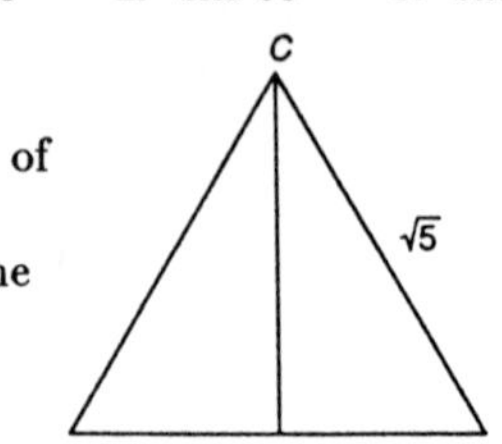


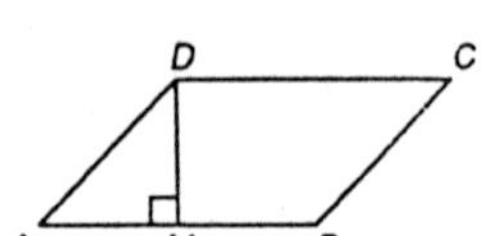


84. In parallelogram $ABCD$ shown in the margin, $AM = MB$, $BC = \sqrt{2}$, and $DC = 2$. What is the area of the parallelogram?

**85.** Use similar triangles and a proportion to find the height of the cliff shown in the figure below.

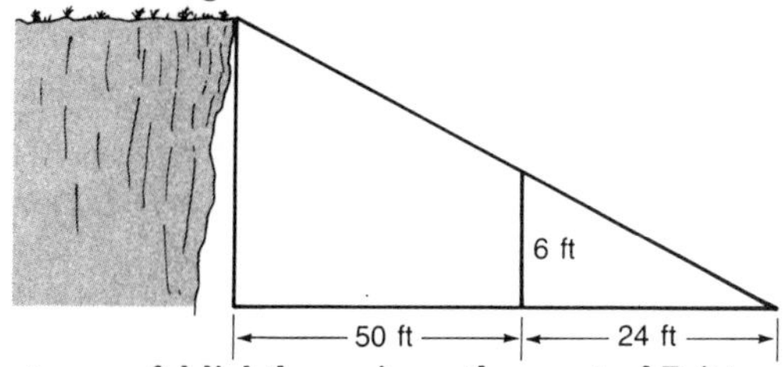


**86.** The most powerful lighthouse is on the coast of Brittany, France, and is about 160 ft tall. Suppose you are in a boat just off the coast, as shown in the figure. Determine your distance (to the nearest foot) from the base of the lighthouse if $m\angle B = 12°$.

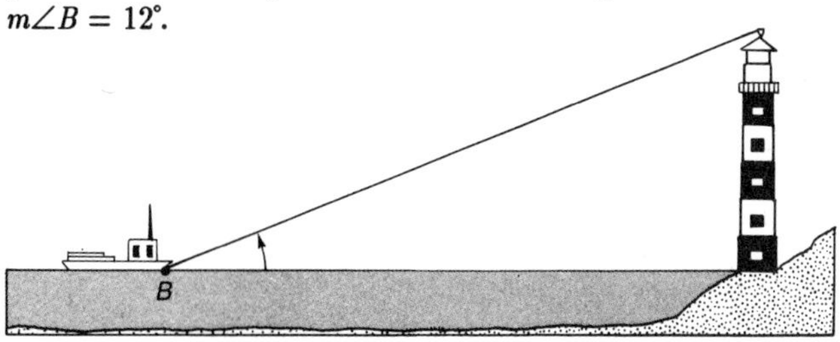


**87.** If a tree casts a shadow of 18 ft at the same time that a yardstick (3 ft) casts a shadow of 2 ft, find the height of the tree (to the nearest foot).

**88.** Use similar triangles to find the length of the sand pit in the accompanying figure.

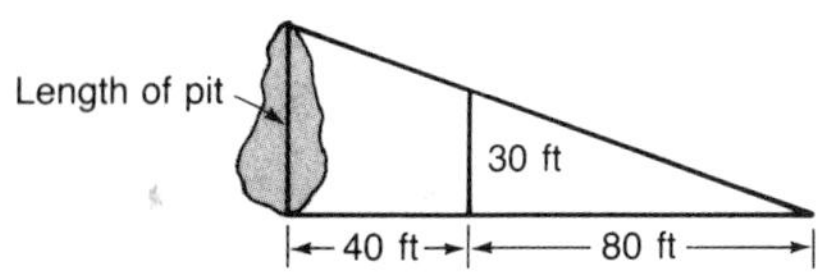


**89.** Find the distance $PA$ (rounded to the nearest foot) across the river, as shown in the following figure, if it is known that distance $DA$ is 50 ft and the measure of the angle at $D$ is 38°.

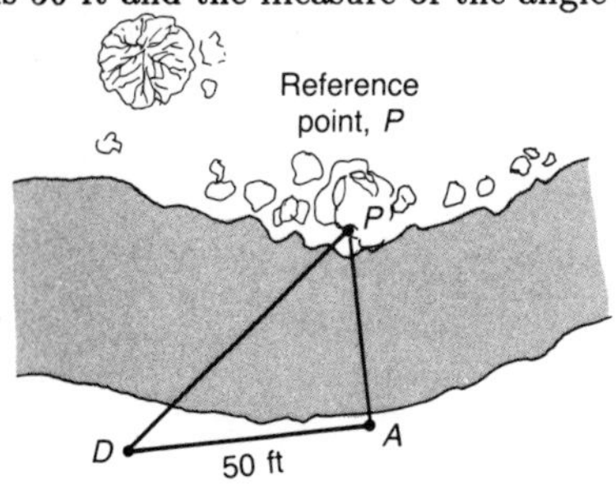


***90.** **a.** In your own words, explain what is meant by inductive reasoning.

**b.** In your own words, explain what is meant by deductive reasoning.

---

*Optional chapter

*91. Does the following story illustrate inductive or deductive reasoning?

*Q:* What has 18 legs and catches flies?
*A:* I don't know, what?
*Q:* A baseball team. What has 36 legs and catches flies?
*A:* I don't known that either.
*Q:* Two baseball teams. If the United States has 100 Senators, and each state has two Senators, what does...
*A:* I know this one!
*Q:* Good. What does each state have?
*A:* Three baseball teams!

*92. Give the next two terms for each of the given sequences.

**a.** 6, 12, 24, 48, 96, ...
**b.** 48, 300, 51, 400, 54, 500, ...
**c.** 1, 10, 100, 1000, ...
**d.** 3, 15, 27, 39, ...
**e.** 8, 8, 8, 8, ...

*Use a Venn diagram to check the validity of the arguments given in problems 93–96.*

*93. All artists are creative.
Some musicians are artists.
Therefore, some musicians are creative.

*94. No apples are bananas.
All apples are fruit.
Therefore, no bananas are fruit.

*95. All birds have wings.
All flies have wings.
Therefore, some flies are birds.

*96. All rectangles are polygons.
All squares are rectangles.
Therefore, all squares are polygons.

*Use the results of the following survey in Problems 97–100. The survey included 500 motorists.*

*140 had received a ticket for speeding.*
*120 had received a ticket for failure to yield the right-of-way.*
*98 had received a ticket for failure to stop at a stop sign.*
*41 had received tickets for speeding and failure to yield.*
*35 had received tickets for speeding and failure to stop at a stop sign.*
*47 had received tickets for failure to yield and failure to stop at a stop sign.*
*24 had received tickets for all three violations.*

*97. How many motorists did not receive any tickets?
*98. How many motorists received only one ticket?
*99. How many motorists received exactly two tickets?
*100. How many motorists received only a ticket for speeding?

*Optional chapter

# APPENDIX A TABLES

TABLE I
Squares and Square Roots

| $n$ | $n^2$ | $\sqrt{n}$ | $\sqrt{10n}$ | $n$ | $n^2$ | $\sqrt{n}$ | $\sqrt{10n}$ |
|---|---|---|---|---|---|---|---|
| 1 | 1 | 1.000 | 3.162 | 51 | 2601 | 7.141 | 22.583 |
| 2 | 4 | 1.414 | 4.472 | 52 | 2704 | 7.211 | 22.804 |
| 3 | 9 | 1.732 | 5.477 | 53 | 2809 | 7.280 | 23.022 |
| 4 | 16 | 2.000 | 6.325 | 54 | 2916 | 7.348 | 23.238 |
| 5 | 25 | 2.236 | 7.071 | 55 | 3025 | 7.416 | 23.452 |
| 6 | 36 | 2.449 | 7.746 | 56 | 3136 | 7.483 | 23.664 |
| 7 | 49 | 2.646 | 8.367 | 57 | 3249 | 7.550 | 23.875 |
| 8 | 64 | 2.828 | 8.944 | 58 | 3364 | 7.616 | 24.083 |
| 9 | 81 | 3.000 | 9.487 | 59 | 3481 | 7.681 | 24.290 |
| 10 | 100 | 3.162 | 10.000 | 60 | 3600 | 7.746 | 24.495 |
| 11 | 121 | 3.317 | 10.488 | 61 | 3721 | 7.810 | 24.698 |
| 12 | 144 | 3.464 | 10.954 | 62 | 3844 | 7.874 | 24.900 |
| 13 | 169 | 3.606 | 11.402 | 63 | 3969 | 7.937 | 25.100 |
| 14 | 196 | 3.742 | 11.832 | 64 | 4096 | 8.000 | 25.298 |
| 15 | 225 | 3.873 | 12.247 | 65 | 4225 | 8.062 | 25.495 |
| 16 | 256 | 4.000 | 12.649 | 66 | 4356 | 8.124 | 25.690 |
| 17 | 289 | 4.123 | 13.038 | 67 | 4489 | 8.185 | 25.884 |
| 18 | 324 | 4.243 | 13.416 | 68 | 4624 | 8.246 | 26.077 |
| 19 | 361 | 4.359 | 13.784 | 69 | 4761 | 8.307 | 26.268 |
| 20 | 400 | 4.472 | 14.142 | 70 | 4900 | 8.367 | 26.458 |
| 21 | 441 | 4.583 | 14.491 | 71 | 5041 | 8.426 | 26.646 |
| 22 | 484 | 4.690 | 14.832 | 72 | 5184 | 8.485 | 26.833 |
| 23 | 529 | 4.796 | 15.166 | 73 | 5329 | 8.544 | 27.019 |
| 24 | 576 | 4.899 | 15.492 | 74 | 5476 | 8.602 | 27.203 |
| 25 | 625 | 5.000 | 15.811 | 75 | 5625 | 8.660 | 27.386 |
| 26 | 676 | 5.099 | 16.125 | 76 | 5776 | 8.718 | 27.568 |
| 27 | 729 | 5.196 | 16.432 | 77 | 5929 | 8.775 | 27.749 |
| 28 | 784 | 5.292 | 16.733 | 78 | 6084 | 8.832 | 27.928 |
| 29 | 841 | 5.385 | 17.029 | 79 | 6241 | 8.888 | 28.107 |
| 30 | 900 | 5.477 | 17.321 | 80 | 6400 | 8.944 | 28.284 |
| 31 | 961 | 5.568 | 17.607 | 81 | 6561 | 9.000 | 28.460 |
| 32 | 1024 | 5.657 | 17.889 | 82 | 6724 | 9.055 | 28.636 |
| 33 | 1089 | 5.745 | 18.166 | 83 | 6889 | 9.110 | 28.810 |
| 34 | 1156 | 5.831 | 18.439 | 84 | 7056 | 9.165 | 28.983 |
| 35 | 1225 | 5.916 | 18.708 | 85 | 7225 | 9.220 | 29.155 |
| 36 | 1296 | 6.000 | 18.974 | 86 | 7396 | 9.274 | 29.326 |
| 37 | 1369 | 6.083 | 19.235 | 87 | 7569 | 9.327 | 29.496 |
| 38 | 1444 | 6.164 | 19.494 | 88 | 7744 | 9.381 | 29.665 |
| 39 | 1521 | 6.245 | 19.748 | 89 | 7921 | 9.434 | 29.833 |
| 40 | 1600 | 6.325 | 20.000 | 90 | 8100 | 9.487 | 30.000 |
| 41 | 1681 | 6.403 | 20.248 | 91 | 8281 | 9.539 | 30.166 |
| 42 | 1764 | 6.481 | 20.494 | 92 | 8464 | 9.592 | 30.332 |
| 43 | 1849 | 6.557 | 20.736 | 93 | 8649 | 9.644 | 30.496 |
| 44 | 1936 | 6.633 | 20.976 | 94 | 8836 | 9.695 | 30.659 |
| 45 | 2025 | 6.708 | 21.213 | 95 | 9025 | 9.747 | 30.822 |
| 46 | 2116 | 6.782 | 21.448 | 96 | 9216 | 9.798 | 30.984 |
| 47 | 2209 | 6.856 | 21.679 | 97 | 9409 | 9.849 | 31.145 |
| 48 | 2304 | 6.928 | 21.909 | 98 | 9604 | 9.899 | 31.305 |
| 49 | 2401 | 7.000 | 22.136 | 99 | 9801 | 9.950 | 31.464 |
| 50 | 2500 | 7.071 | 22.361 | 100 | 10000 | 10.000 | 31.623 |

TABLE II
Trigonometric Functions

| Degrees | sin x | cos x | tan x | Degrees | sin x | cos x | tan x |
|---|---|---|---|---|---|---|---|
| 1 | 0.0175 | 0.9998 | 0.0175 | 46 | 0.7193 | 0.6947 | 1.0355 |
| 2 | 0.0349 | 0.9994 | 0.0349 | 47 | 0.7314 | 0.6820 | 1.0724 |
| 3 | 0.0523 | 0.9986 | 0.0524 | 48 | 0.7431 | 0.6691 | 1.1106 |
| 4 | 0.0698 | 0.9976 | 0.0699 | 49 | 0.7547 | 0.6561 | 1.1504 |
| 5 | 0.0872 | 0.9962 | 0.0875 | 50 | 0.7660 | 0.6428 | 1.1918 |
| 6 | 0.1045 | 0.9945 | 0.1051 | 51 | 0.7771 | 0.6293 | 1.2349 |
| 7 | 0.1219 | 0.9925 | 0.1228 | 52 | 0.7880 | 0.6157 | 1.2799 |
| 8 | 0.1392 | 0.9903 | 0.1405 | 53 | 0.7986 | 0.6018 | 1.3270 |
| 9 | 0.1564 | 0.9877 | 0.1584 | 54 | 0.8090 | 0.5878 | 1.3764 |
| 10 | 0.1736 | 0.9848 | 0.1763 | 55 | 0.8192 | 0.5736 | 1.4281 |
| 11 | 0.1908 | 0.9816 | 0.1944 | 56 | 0.8290 | 0.5592 | 1.4826 |
| 12 | 0.2079 | 0.9781 | 0.2126 | 57 | 0.8387 | 0.5446 | 1.5399 |
| 13 | 0.2250 | 0.9744 | 0.2309 | 58 | 0.8480 | 0.5299 | 1.6003 |
| 14 | 0.2419 | 0.9703 | 0.2493 | 59 | 0.8572 | 0.5150 | 1.6643 |
| 15 | 0.2588 | 0.9659 | 0.2679 | 60 | 0.8660 | 0.5000 | 1.7321 |
| 16 | 0.2756 | 0.9613 | 0.2867 | 61 | 0.8746 | 0.4848 | 1.8040 |
| 17 | 0.2924 | 0.9563 | 0.3057 | 62 | 0.8829 | 0.4695 | 1.8807 |
| 18 | 0.3090 | 0.9511 | 0.3249 | 63 | 0.8910 | 0.4540 | 1.9626 |
| 19 | 0.3256 | 0.9455 | 0.3443 | 64 | 0.8988 | 0.4384 | 2.0503 |
| 20 | 0.3420 | 0.9397 | 0.3640 | 65 | 0.9063 | 0.4226 | 2.1445 |
| 21 | 0.3584 | 0.9336 | 0.3839 | 66 | 0.9135 | 0.4067 | 2.2460 |
| 22 | 0.3746 | 0.9272 | 0.4040 | 67 | 0.9205 | 0.3907 | 2.3559 |
| 23 | 0.3907 | 0.9205 | 0.4245 | 68 | 0.9272 | 0.3746 | 2.4751 |
| 24 | 0.4067 | 0.9135 | 0.4452 | 69 | 0.9336 | 0.3584 | 2.6051 |
| 25 | 0.4226 | 0.9063 | 0.4663 | 70 | 0.9397 | 0.3420 | 2.7475 |
| 26 | 0.4384 | 0.8988 | 0.4877 | 71 | 0.9455 | 0.3256 | 2.9042 |
| 27 | 0.4540 | 0.8910 | 0.5095 | 72 | 0.9511 | 0.3090 | 3.0777 |
| 28 | 0.4695 | 0.8829 | 0.5317 | 73 | 0.9563 | 0.2924 | 3.2709 |
| 29 | 0.4848 | 0.8746 | 0.5543 | 74 | 0.9613 | 0.2756 | 3.4874 |
| 30 | 0.5000 | 0.8660 | 0.5774 | 75 | 0.9659 | 0.2588 | 3.7321 |
| 31 | 0.5150 | 0.8572 | 0.6009 | 76 | 0.9703 | 0.2419 | 4.0108 |
| 32 | 0.5299 | 0.8480 | 0.6249 | 77 | 0.9744 | 0.2250 | 4.3315 |
| 33 | 0.5446 | 0.8387 | 0.6494 | 78 | 0.9781 | 0.2079 | 4.7046 |
| 34 | 0.5592 | 0.8290 | 0.6745 | 79 | 0.9816 | 0.1908 | 5.1446 |
| 35 | 0.5736 | 0.8192 | 0.7002 | 80 | 0.9848 | 0.1736 | 5.6713 |
| 36 | 0.5878 | 0.8090 | 0.7265 | 81 | 0.9877 | 0.1564 | 6.3138 |
| 37 | 0.6018 | 0.7986 | 0.7536 | 82 | 0.9903 | 0.1392 | 7.1154 |
| 38 | 0.6157 | 0.7880 | 0.7813 | 83 | 0.9925 | 0.1219 | 8.1444 |
| 39 | 0.6293 | 0.7771 | 0.8098 | 84 | 0.9945 | 0.1045 | 9.5144 |
| 40 | 0.6428 | 0.7660 | 0.8391 | 85 | 0.9962 | 0.0872 | 11.4300 |
| 41 | 0.6561 | 0.7547 | 0.8693 | 86 | 0.9976 | 0.0698 | 14.3007 |
| 42 | 0.6691 | 0.7431 | 0.9004 | 87 | 0.9986 | 0.0523 | 19.0812 |
| 43 | 0.6820 | 0.7314 | 0.9325 | 88 | 0.9994 | 0.0349 | 28.6362 |
| 44 | 0.6947 | 0.7193 | 0.9657 | 89 | 0.9998 | 0.0175 | 57.2898 |
| 45 | 0.7071 | 0.7071 | 1.0000 | 90 | 1.0000 | 0.0000 | undefined |

# APPENDIX B ANSWERS

**Problem Set 1**

**1.** quadrilateral **2.** pentagon **3.** octagon **4.** decagon **5.** triangle **6.** hexagon **7.** octagon **8.** dodecagon **9.** quadrilateral **10.** heptagon **11.** octagon **12.** nonagon **13. a.** obtuse **b.** obtuse **c.** acute **d.** acute **e.** straight **14. a.** right **b.** acute **c.** straight **d.** acute **e.** acute **15.** answers vary; $m\angle COB = m\angle DOA$ or $m\angle DOC = m\angle AOB$ **16.** $\angle AOC$ **17. a.** obtuse **b.** obtuse **c.** acute **d.** acute **e.** straight **18. a.** right **b.** acute **c.** straight **d.** acute **e.** acute **19.** answers vary; $m\angle COB = m\angle DOA$ or $m\angle DOC = m\angle AOB$ **20.** $\angle AOC$ **21.** 30° **22.** 90° **23.** 75° **24.** 90° **25.** 40° **26.** 80° **27.** vertical **28.** supplementary **29.** alternate interior **30.** corresponding **31.** alternate exterior **32.** alternate interior **33.** F **34.** F **35.** T **36.** F **37.** T **38.** F **39.** $m\angle 1 = m\angle 3 = m\angle 5 = m\angle 7 = 110°$; $m\angle 2 = m\angle 4 = m\angle 6 = m\angle 8 = 70°$ **40.** $m\angle 1 = m\angle 3 = m\angle 5 = m\angle 7 = 115°$; $m\angle 2 = m\angle 4 = m\angle 6 = m\angle 8 = 65°$ **41.** $m\angle 1 = m\angle 3 = m\angle 5 = m\angle 7 = 161°$; $m\angle 2 = m\angle 4 = m\angle 6 = m\angle 8 = 19°$ **42.** $m\angle 1 = m\angle 3 = m\angle 5 = m\angle 7 = 153°$; $m\angle 2 = m\angle 4 = m\angle 6 = m\angle 8 = 27°$ **43.** 112° **44.** 120° **45.** 60° **46.** 50° **47.** 88° **48.** 62° **49.** 145° **50.** 35° **51.** 56° **52.** 49° **53.** 75° **54.** 95° **55.** 80° **56.** 95° **57.** 100° **58.** 115° **59.** 20° **60.** 10° **61.** 21° **62.** 12.5° **63.** 6° **64.** 5° **65.** 13°; 53°; 114° **66.** 25°; 65°; 90°

**Problem Set 2**

**1.** 18 in. **2.** 30 ft **3.** 150 m **4.** 24 cm **5.** 40 mm **6.** 960 ft **7.** 18 yd **8.** 9 dm **9.** 390 m **10.** 21.99 m **11.** 31.42 in. **12.** 157.08 ft **13.** 15.08 m **14.** 502.65 ft **15.** 75.40 in. **16.** 46.0 ft **17.** 52.0 in. **18.** 530.0 cm **19.** 7.3 in. **20.** 20.0 cm **21.** 15.2 cm **22.** 257.1 cm **23.** 35.7 in. **24.** 64.3 ft **25.** 397.1 ft **26.** 11.1 cm **27.** 11.1 cm **28.** 23 ft **29.** 6 in. **30.** 48.5 cm **31.** 18.3 mm **32.** 65 ft **33.** 300 m **34.** The sides are 26 in., 39 in., and 52 in. **35.** The sides are 40 cm, 60 cm, 60 cm, 40 cm, and 80 cm.

**36.**

$$2r + \tfrac{1}{4}C = 4 + \pi$$
$$2r + \tfrac{1}{4}(2r\pi) = 4 + \pi$$
$$2r + \tfrac{1}{2}r\pi = 4 + \pi$$
$$4r + r\pi = 2(4 + \pi)$$
$$r(4 + \pi) = 2(4 + \pi)$$
$$r = 2$$

Thus, the area is $\frac{1}{4}A = \frac{1}{4}\pi(2)^2 = \pi$.

**37.** Let $s$ be the length of one side of the triangle.

$$3s = 4$$
$$s = \tfrac{4}{3}$$

Then $P = 6s = 6(\frac{4}{3}) = 8$.

**38.** 66 dm **39.** 8 cm by 19 cm

**Problem Set 3**
**1.** 15 in.$^2$ **2.** 12 ft$^2$ **3.** 1,196 m$^2$ **4.** 36 mi$^2$ **5.** 100 mm$^2$ **6.** 250 in.$^2$ **7.** 7,560 ft$^2$ **8.** 13.5 m$^2$ **9.** 136.5 dm$^2$ **10.** 40 ft$^2$ **11.** 6,600 cm$^2$ **12.** 55 ft$^2$ **13.** 28 in.$^2$ **14.** 145 mm$^2$ **15.** 314.2 in.$^2$ **16.** 706.9 in.$^2$ **17.** 78.5 in.$^2$ **18.** 307.9 in.$^2$ **19.** 19.6 dm$^2$ **20.** 353.4 cm$^2$ **21.** 7.6 cm$^2$ **22.** 4.4 cm$^2$ **23.** 320 in.$^2$ **24.** 46 ft$^2$ **25.** 520 mm$^2$ **26.** 6 yd$^2$ **27.** 1,546 cm$^2$ **28.** 1,407 cm$^2$ **29.** 251 in.$^2$ **30.** 42 in.$^2$ **31.** $4a^2(9 - \pi)$ **32.** $4b^2(\pi - 1)$ **33.** $5s^2$ **34.** $7s^2$ **35.** $a^2(6 + \frac{1}{2}\pi)$ or $\frac{1}{2}a^2(12 + \pi)$

**36.** $b^2(6 - \frac{1}{2}\pi)$ or $\frac{1}{2}b^2(12 - \pi)$ **37.** A **38.** C **39.** 22.2 acres **40.** 196,020 ft$^2$

**41.** 6 cm$^2$ **42.** $\frac{8}{3}$ **43.** $20\sqrt{2}$ ft

**44.** Label $x$ and $y$:

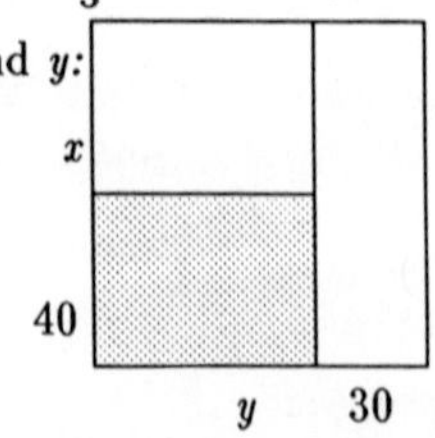


Since it is a square,

$$x + 40 = y + 30$$
$$y = x + 10$$

Thus,

$$(x + 40)^2 = 6{,}400$$
$$x + 40 = 80$$
$$x = 40$$

So $y = 50$, and the desired area is $40(50) = 2{,}000$ ft$^2$.

**45.** \$1,664 **46.** \$117 **47.** \$6,460 **48.** A small pizza is about 79 in.$^2$, which is approximately 10.1 cents per in.$^2$; a large pizza is about 154 in.$^2$, which is approximately 9.0 cents per in.$^2$; the large pizza is the better buy. **49.** $6\frac{1}{8} \times 2\frac{5}{8} = \frac{1{,}029}{64} \approx 16$ in.$^2$ **50.** $2\frac{1}{8} \times 3\frac{3}{8} = \frac{459}{64} \approx 7$ in.$^2$ **51.** $16\pi$ **52.** It is quadrupled. **53.** 19 in.$^2$ **54.** 62 in.

**Problem Set 4**
**1.** 60 cm$^3$ **2.** 80 cm$^3$ **3.** 42 cm$^3$ **4.** 125 ft$^3$ **5.** 1,728 in.$^3$ **6.** 8,000 cm$^3$ **7.** 42 m$^3$ **8.** 24 ft.$^3$ **9.** 400 mm$^3$ **10.** 3,600 in.$^3$ **11.** 96,000 cm$^3$ **12.** 51.45 cm$^3$ **13.** 30 ft$^3$ **14.** 14,137 cm$^3$ **15.** 19 cm$^3$ **16.** 30 in.$^3$ **17.** 600 cm$^3$ **18.** 64 cm$^3$ **19.** 47 in.$^3$ **20. a.** 45.375 ft$^3$ **b.** 26.375 ft$^3$ **21.** No; the exterior dimensions are 24 ft$^3$, so it can't be 27 ft$^3$ inside. **22.** $6 \times 2 \times 2 + 8 \times 2 \times 2 = 56$ **23.** 2.5 yd$^3$ **24.** $9\frac{1}{2}$ yd$^3$ **25.** $5\frac{1}{2}$ yd$^3$ **26.** $11\frac{1}{2}$ yd$^3$ **27.** $1\frac{1}{2}$ yd$^3$ **28.** $3\frac{1}{2}$ yd$^3$

**Problem Set 5**
**1. a.** $RI = NG$ and $TI = AG$ **b.** $RI = NG$ **c.** $\angle T \simeq \angle A$ **2. a.** $BC = EF$ and $AB = DE$ **b.** $AB = DE$ **c.** $\angle C \simeq \angle F$ **3.** $AB = ED$; $AC = EF$; $CB = FD$; $\angle A \simeq \angle E$; $\angle B \simeq \angle D$; $\angle C \simeq \angle F$ **4.** $GH = G'H'$; $GI = G'I'$; $HI = H'I'$; $\angle G \simeq \angle G'$; $\angle H \simeq \angle H'$; $\angle I \simeq \angle I'$
**5.** $RS = TU$; $RT = RT$; $ST = UR$; $\angle SRT \simeq \angle URT$; $\angle S \simeq \angle U$; $\angle STR \simeq \angle URT$
**6.** $WX = YX$; $WZ = YZ$; $XZ = XZ$; $\angle W \simeq \angle Y$; $\angle WXZ \simeq \angle YZX$; $\angle WZX \simeq \angle YXZ$
**7.** $AB = DE$; $BC = EF$; $AC = DF$; $\angle A \simeq \angle D$; $\angle B \simeq \angle E$; $\angle C \simeq \angle F$ **8.** $RE = TA$; $EC = AN$; $RC = TN$; $\angle R \simeq \angle T$; $\angle E \simeq \angle A$; $\angle C \simeq \angle N$ **9.** Since these figures are similar, the ratios of the lengths of their sides are proportional; thus,

$$\frac{DE}{DA} = \frac{EF}{AB}$$
$$\frac{4}{24} = \frac{EF}{12}$$
$$EF = 2$$

**10.** $m\angle D = 40°$, since corresponding angles have the same measure in similar figures.
**11.** congruent; ASA **12.** congruent; SAS **13.** congruent; SAS **14.** congruent; ASA
**15.** similar; $AB \simeq ED$; $AC \simeq EF$; $CB \simeq FD$; $\angle A \simeq \angle E$; $\angle B \simeq \angle D$; $\angle C \simeq \angle F$
**16.** not similar **17.** not similar **18.** similar; $GH \simeq JK$; $GI \simeq JL$; $IH \simeq LK$; $\angle G \simeq \angle J$; $\angle H \simeq \angle K$; $\angle I \simeq \angle L$ **19.** similar; $MN \simeq SQ$; $MP \simeq SR$; $PN \simeq RQ$; $\angle M \simeq \angle S$; $\angle P \simeq \angle R$; $\angle N \simeq \angle Q$ **20.** similar; $TV \simeq WY$; $TU \simeq WX$; $VU \simeq YX$; $\angle T \simeq \angle W$; $\angle U \simeq \angle X$; $\angle V \simeq \angle Y$ **21.** 4 **22.** 6 **23.** $\frac{16}{3}$ **24.** $\frac{40}{7}$ **25.** $\frac{8}{3}$

**26.** $\frac{10}{9}$ **27.** $5\sqrt{2}$ **28.** $\frac{8}{3}\sqrt{3}$ **29.** 6 **30.** 20 **31.** $\frac{7}{2}\sqrt{2}$ **32.** $\frac{15}{2}$ **33.** $\frac{9}{2}\sqrt{2}$ **34.** $6\sqrt{3}$

**35.** $\frac{20}{3}\sqrt{3}$ **36.** $14\sqrt{6}$ **37.** .8290 **38.** .2588 **39.** .8746 **40.** .3090 **41.** .5878

**42.** .9903 **43.** 0 **44.** .8290 **45.** .4452 **46.** 1.2799 **47.** 3.7321 **48.** 57.2900

**49.** $\sin A = \frac{5}{13} = .3846$; $\cos A = \frac{12}{13} = .9231$; $\tan A = \frac{5}{12} = .4167$ **50.** $\sin A = \frac{12}{13} = .9231$; $\cos A = \frac{5}{13} = .3846$; $\tan A = \frac{12}{5} = 2.4$ **51.** $\sin A = \frac{3}{5} = .6$; $\cos A = \frac{4}{5} = .8$; $\tan A = \frac{3}{4} = .75$

**52.** $\sin A = \frac{\sqrt{35}}{6} = .9860$; $\cos A = \frac{1}{6} = .1667$; $\tan A = \frac{\sqrt{35}}{1} = 5.9161$ **53.** $\sin A = \frac{1}{6} = .1667$; $\cos A = \frac{\sqrt{35}}{6} = .9860$; $\tan A = \frac{1}{\sqrt{35}} = .1690$ **54.** $\sin A = \frac{1}{\sqrt{5}} = .4472$; $\cos A = \frac{2}{\sqrt{5}} = .8944$; $\tan A = \frac{1}{2} = .5000$ **55.** $\sin A = \frac{1}{2} = .5000$; $\cos A = \frac{\sqrt{3}}{2} = .8660$; $\tan A = \frac{1}{\sqrt{3}} = .5774$

**56.** $\sin A = \frac{6}{\sqrt{37}} = .9864$; $\cos A = \frac{1}{\sqrt{37}} = .1644$; $\tan A = 6 = 6.0000$ **57.** $\sin A = \frac{\sqrt{2}}{3} = .4714$; $\cos A = \frac{\sqrt{7}}{3} = .8819$; $\tan A = \frac{\sqrt{2}}{\sqrt{7}} = .5345$ **58.** 24 ft **59.** 25 ft **60.** 10 ft **61.** $2\sqrt{2}$ in.

**62.** $3\sqrt{2}$ ft **63.** 200 ft **64.** $5\sqrt{13}$ ft; 18 ft; 55 ft **65.** $CD = \sqrt{2^2 - 1^2} = \sqrt{3}$

**66.** Since $\triangle ABC \sim \triangle DEC$, $\frac{4}{AB} = \frac{3}{5}$; thus, $AB = \frac{20}{3}$ **67.** 260 ft **68.** 125 ft **69.** 24 ft

**70.** 29 ft **71.** 70 ft **72.** 109 ft

**Problem Set 6**

**1.** Answers vary **2.** Answers vary **3.** Answers vary; a type of logical reasoning that has a form consisting of two premises and a conclusion. **4.** 17, 20 (common difference, 3) **5.** 1, 2 (number of 1s increase by one, separated by 2s) **6.** 22, 29 (add 1, add 2, add 3, ...) **7.** 47, 76 (add two terms for next term; 1 + 3 = 4; 3 + 4 = 7; ... 18 + 29 = 47; 29 + 47 = 76) **8.** 96, 192 (common ratio, 2) **9.** 1215, 3645 (common ratio, 3) **10.** 85, 79 (subtract 1, subtract 2, subtract 3, ...) **11.** 36, 49 (squares of consecutive counting numbers) **12.** 2, 5 (number of 5s between 2s increases by 1) **13.** 45, 55 (common difference, 10) **14.** N, E (alternating letters of the alphabet and a pattern where one letter, then two, then three letters skipped) **15.** L, M (listing of letters that do not contain curves) **16.** (1) football players who are not basketball players and also are not baseball players; (2) basketball players who are not football players and also are not baseball players; (3) baseball players who are not football players and also are not basketball players; (4) persons who play both football and baseball, but not basketball; (5) persons who play both football and basketball, but not baseball; (6) persons who play basketball and baseball, but not football; (7) persons who play baseball, football, and basketball; (8) persons who do not play baseball, football, or basketball. **17.** (1) equilateral figures that are not equiangular and also are not quadrilaterals; (2) equiangular figures that are not equilateral and also are not quadrilaterals; (3) quadrilaterals that are not equilateral and also are not equiangular; (4) figures that are both equilateral and quadrilaterals, but are not equiangular; (5) figures that are both equilateral and equiangular, but are not quadrilaterals; (6) figures are both quadrilaterals and equiangular, but are not equilateral; (7) figures that are equilateral, equiangular, and quadrilaterals; (8) figures that are not equilateral, equiangular, or quadrilaterals. **18.** valid **19.** valid **20.** valid **21.** valid **22.** The set of men is the complement of the set of women; not valid. **23.** not valid **24.** not valid **25.** valid **26.** not valid **27.** valid **28.** valid **29.** 23 in $A$ only; 25 in $B$ only; 14 in $C$ only; 13 in $A$ and $B$ but not in $C$; 8 in $A$ and $C$ but not in $B$; 2 in $B$ and $C$ but not in $A$; 11 in all three; and 4 in none of the three **30.** No, there were 102 polled, not 100.

**Problem Set 7**

**1.** quadrilateral **2.** acute **3.** vertical; corresponding; alternate; alternate **4.** supplementary **5.** perpendicular **6.** 180° **7.** $P = 3s$ **8.** $P = 2l + 2w$ **9.** $C = \pi d$ **10.** $d = 2r$ **11.** $A = lw$ **12.** $A = s^2$ **13.** 43,560 **14.** $A = bh$ **15.** $A = \frac{1}{2}h(b + B)$ **16.** $A = \pi r^2$ **17.** add; faces **18.** $V = s^3$ **19.** $V = lwh$ **20.** $V = Bh$ **21.** $V = \frac{4}{3}\pi r^3$ **22.** $V = \frac{1}{3}Bh$ **23.** divide by 27 **24.** congruent **25.** *SSS, SAS, ASA* **26.** corresponding sides are proportional **27.** $c^2 = a^2 + b^2$ for a right triangle with hypotenuse $c$ **28.** $\frac{\text{SIDE OPPOSITE } B}{\text{HYPOTENUSE}}$; $\frac{\text{SIDE ADJACENT } B}{\text{HYPOTENUSE}}$; $\frac{\text{SIDE OPPOSITE } B}{\text{SIDE ADJACENT } B}$ **29.** looking for a pattern and making observations **30.** using a formal structure based on a set of undefined terms and a set of accepted unproved axioms or premises. **31.** premises; conclusion **32. a.** hexagon **b.** quadrilateral **c.** octagon **d.** triangle **e.** pentagon **33. a.** pentagon **b.** quadrilateral **c.** hexagon **34. a.** acute **b.** acute **c.** straight **d.** right **e.** obtuse **35.** *Reasons may vary* **a.** 50°; vertical angles **b.** 130°; vertical angles **c.** 90°; alternate interior angles **d.** 50°; vertical angles **e.** 130°; supplementary angles **36.** $m\angle A = 85°$ **37.** $x = 25$; 45°, 50°, 85° **38. a.** 128° **b.** 62° **c.** 75° **d.** 30° **39.** 28 ft **40.** 60 in. **41.** 28 m **42.** 30 ft **43.** 40.84 cm **44.** 71.42 ft **45.** 16 cm **46.** 43.14 in. **47.** 132.73 cm$^2$ **48.** 157.08 ft$^2$ **49.** 6 cm$^2$ **50.** 121.81 in.$^2$ **51.** $P \approx 43$ in.; $A \approx 111$ in.$^2$ **52.** $25\pi$ **53.** 13.5 in.$^2$ **54.** 12 cm **55.** 504 in.$^2$ **56.** 743 in.$^2$ **57.** 464 dm$^2$ **58.** 48 cm$^2$ **59.** 392 in.$^3$ **60.** 1,848 in.$^3$ **61.** 896 dm$^3$ **62.** 20 cm$^3$ **63.** 28 cm$^3$ **64.** 17 cm$^3$ **65.** 34 m$^3$ **66.** 42 cm$^3$ **67.** 4.2 ft$^3$ **68.** 21 yd$^2$ **69.** square, \$400; circular, \$314 **70.** $6\frac{1}{2}$ yd$^3$ **71.** 420 in.$^3$ **72.** $3\frac{1}{2}$ yd$^3$ **73.** 524 ft$^3$ **74.** multiplied by 8 **75.** $TI = AI$; $TR = AR$; $RI = RI$; $\angle TRI \simeq \angle ARI$; $\angle TIR \simeq \angle AIR$; $\angle RTI \simeq \angle RAI$ **76. a.** 51° **b.** 129° **c.** $AI = 6 \tan 39° \approx 4.8587$; the length is about 5 in. **d.** Yes; the triangles are congruent, so they must be similar. **77.** 18 **78.** $\sqrt{21}$ **79. a.** 6 **b.** $\frac{8}{3}$ **c.** $2\sqrt{13} \approx 7.2$ **d.** 4.3 **80. a.** $\tan 35° = \frac{45}{b}$; $b \approx 64$ **b.** $\tan 22° = \frac{b}{38}$; $b \approx 15$ **c.** $\sin 17° = \frac{19}{c}$; $c \approx 65$ **d.** $\cos 28° = \frac{128}{c}$; $c \approx 144.97$ **81. a.** .8572 **b.** 0 **c.** .4540 **d.** 7.1154 **e.** 1 **82. a.** 4.3315 **b.** .1219 **c.** .9511 **d.** .9336 **e.** 0 **83.** $\frac{\sqrt{15}}{2}$ **84.** $DC = 2$ and $AB = 2$, so $AM = 1$; and since $BC = \sqrt{2}$, $AD = \sqrt{2}$. Use the Pythagorean Theorem on $\triangle ADM$ to find $DM = 1$. Area $= bh = 2(1) = 2$. **85.** 18.5 ft **86.** 753 ft **87.** The tree is 27 ft tall. **88.** The pit is 45 ft long. **89.** The distance across the river is 39 ft. **90. a.** answers vary **b.** answers vary **91.** inductive reasoning **92. a.** 192, 384 (common ratio, 2) **b.** 57, 600 (alternating; add 3, then add 100) **c.** 10,000, 100,000 (common ratio, 10) **d.** 51, 63 (common difference, 12) **e.** 8, 8 (common difference, 0; or common ratio, 1) **93.** valid **94.** not valid **95.** not valid **96.** valid **97.** 241 **98.** 184 **99.** 51 **100.** 88

# INDEX